Functional Genomics

METHODS IN MOLECULAR BIOLOGY™

John M. Walker, SERIES EDITOR

236. **Plant Functional Genomics:** *Methods and Protocols*, edited by *Erich Grotewold, 2003*
235. ***E. coli* Plasmid Vectors:** *Methods and Applications*, edited by *Nicola Casali and Andrew Preston, 2003*
234. **p53 Protocols**, edited by *Sumitra Deb and Swati Palit Deb, 2003*
233. **Protein Kinase C Protocols**, edited by *Alexandra C. Newton, 2003*
232. **Protein Misfolding and Disease:** *Principles and Protocols*, edited by *Peter Bross and Niels Gregersen, 2003*
231. **Directed Evolution Library Creation:** *Methods and Protocols*, edited by *Frances H. Arnold and George Georgiou, 2003*
230. **Directed Enzyme Evolution:** *Screening and Selection Methods*, edited by *Frances H. Arnold and George Georgiou, 2003*
229. **Lentivirus Gene Engineering Protocols**, edited by *Maurizio Federico, 2003*
228. **Membrane Protein Protocols**: *Expression, Purification, and Characterization*, edited by *Barry S. Selinsky, 2003*
227. **Membrane Transporters:** *Methods and Protocols*, edited by *Qing Yan, 2003*
226. **PCR Protocols, Second Edition,** edited by *John M. S. Bartlett and David Stirling, 2003*
225. **Inflammation Protocols,** edited by *Paul G. Winyard and Derek A. Willoughby, 2003*
224. **Functional Genomics:** *Methods and Protocols*, edited by *Michael J. Brownstein and Arkady B. Khodursky, 2003*
223. **Tumor Suppressor Genes:** *Volume 2: Regulation, Function, and Medicinal Applications*, edited by *Wafik S. El-Deiry, 2003*
222. **Tumor Suppressor Genes:** *Volume 1: Pathways and Isolation Strategies*, edited by *Wafik S. El-Deiry, 2003*
221. **Generation of cDNA Libraries:** *Methods and Protocols*, edited by *Shao-Yao Ying, 2003*
220. **Cancer Cytogenetics:** *Methods and Protocols*, edited by *John Swansbury, 2003*
219. **Cardiac Cell and Gene Transfer:** *Principles, Protocols, and Applications*, edited by *Joseph M. Metzger, 2003*
218. **Cancer Cell Signaling:** *Methods and Protocols*, edited by *David M. Terrian, 2003*
217. **Neurogenetics:** *Methods and Protocols*, edited by *Nicholas T. Potter, 2003*
216. **PCR Detection of Microbial Pathogens:** *Methods and Protocols*, edited by *Konrad Sachse and Joachim Frey, 2003*
215. **Cytokines and Colony Stimulating Factors:** *Methods and Protocols*, edited by *Dieter Körholz and Wieland Kiess, 2003*
214. **Superantigen Protocols,** edited by *Teresa Krakauer, 2003*
213. **Capillary Electrophoresis of Carbohydrates,** edited by *Pierre Thibault and Susumu Honda, 2003*
212. **Single Nucleotide Polymorphisms:** *Methods and Protocols*, edited by *Pui-Yan Kwok, 2003*
211. **Protein Sequencing Protocols, 2nd ed.,** edited by *Bryan John Smith, 2003*
210. **MHC Protocols,** edited by *Stephen H. Powis and Robert W. Vaughan, 2003*
209. **Transgenic Mouse Methods and Protocols,** edited by *Marten Hofker and Jan van Deursen, 2003*
208. **Peptide Nucleic Acids:** *Methods and Protocols*, edited by *Peter E. Nielsen, 2002*
207. **Recombinant Antibodies for Cancer Therapy:** *Methods and Protocols*. edited by *Martin Welschof and Jürgen Krauss, 2002*
206. **Endothelin Protocols,** edited by *Janet J. Maguire and Anthony P. Davenport, 2002*
205. ***E. coli* Gene Expression Protocols,** edited by *Peter E. Vaillancourt, 2002*
204. **Molecular Cytogenetics:** *Protocols and Applications*, edited by *Yao-Shan Fan, 2002*
203. ***In Situ* Detection of DNA Damage:** *Methods and Protocols*, edited by *Vladimir V. Didenko, 2002*
202. **Thyroid Hormone Receptors:** *Methods and Protocols*, edited by *Aria Baniahmad, 2002*
201. **Combinatorial Library Methods and Protocols,** edited by *Lisa B. English, 2002*
200. **DNA Methylation Protocols,** edited by *Ken I. Mills and Bernie H, Ramsahoye, 2002*
199. **Liposome Methods and Protocols,** edited by *Subhash C. Basu and Manju Basu, 2002*
198. **Neural Stem Cells:** *Methods and Protocols*, edited by *Tanja Zigova, Juan R. Sanchez-Ramos, and Paul R. Sanberg, 2002*
197. **Mitochondrial DNA:** *Methods and Protocols*, edited by *William C. Copeland, 2002*
196. **Oxidants and Antioxidants:** *Ultrastructure and Molecular Biology Protocols*, edited by *Donald Armstrong, 2002*
195. **Quantitative Trait Loci:** *Methods and Protocols*, edited by *Nicola J. Camp and Angela Cox, 2002*
194. **Posttranslational Modifications of Proteins:** *Tools for Functional Proteomics*, edited by *Christoph Kannicht, 2002*
193. **RT-PCR Protocols,** edited by *Joe O'Connell, 2002*
192. **PCR Cloning Protocols, 2nd ed.,** edited by *Bing-Yuan Chen and Harry W. Janes, 2002*
191. **Telomeres and Telomerase:** *Methods and Protocols*, edited by *John A. Double and Michael J. Thompson, 2002*
190. **High Throughput Screening:** *Methods and Protocols*, edited by *William P. Janzen, 2002*
189. **GTPase Protocols:** *The RAS Superfamily*, edited by *Edward J. Manser and Thomas Leung, 2002*
188. **Epithelial Cell Culture Protocols,** edited by *Clare Wise, 2002*
187. **PCR Mutation Detection Protocols,** edited by *Bimal D. M. Theophilus and Ralph Rapley, 2002*
186. **Oxidative Stress Biomarkers and Antioxidant Protocols,** edited by *Donald Armstrong, 2002*
185. **Embryonic Stem Cells:** *Methods and Protocols*, edited by *Kursad Turksen, 2002*
184. **Biostatistical Methods,** edited by *Stephen W. Looney, 2002*
183. **Green Fluorescent Protein:** *Applications and Protocols*, edited by *Barry W. Hicks, 2002*
182. ***In Vitro* Mutagenesis Protocols, 2nd ed.**, edited by *Jeff Braman, 2002*
181. **Genomic Imprinting:** *Methods and Protocols*, edited by *Andrew Ward, 2002*
180. **Transgenesis Techniques, 2nd ed.:** *Principles and Protocols*, edited by *Alan R. Clarke, 2002*

METHODS IN MOLECULAR BIOLOGY™

Functional Genomics
Methods and Protocols

Edited by

Michael J. Brownstein
*Laboratory of Genetics, NIMH/NHGRI
National Institutes of Health, Rockville, MD*

and

Arkady B. Khodursky
*Department of Biochemistry, Molecular Biology and Biophysics
University of Minnesota, St. Paul, MN*

Humana Press   **Totowa, New Jersey**

© 2003 Humana Press Inc.
999 Riverview Drive, Suite 208
Totowa, New Jersey 07512

www.humanapress.com

All rights reserved. No part of this book may be reproduced, stored in a retrieval system, or transmitted in any form or by any means, electronic, mechanical, photocopying, microfilming, recording, or otherwise without written permission from the Publisher. Methods in Molecular Biology™ is a trademark of The Humana Press Inc.

The content and opinions expressed in this book are the sole work of the authors and editors, who have warranted due diligence in the creation and issuance of their work. The publisher, editors, and authors are not responsible for errors or omissions or for any consequences arising from the information or opinions presented in this book and make no warranty, express or implied, with respect to its contents.

This publication is printed on acid-free paper. ∞
ANSI Z39.48-1984 (American Standards Institute)

Permanence of Paper for Printed Library Materials.

Cover illustration: 6000-element whole-genome *Escherichia coli* DNA microarray. Image has been manipulated to appear as the cover background. Image courtesy of Arkady Khodursky.

Cover design by Patricia F. Cleary.

For additional copies, pricing for bulk purchases, and/or information about other Humana titles, contact Humana at the above address or at any of the following numbers: Tel.: 973-256-1699; Fax: 973-256-8341; E-mail: humana@humanapr.com; or visit our Website: www.humanapress.com

Photocopy Authorization Policy:
Authorization to photocopy items for internal or personal use, or the internal or personal use of specific clients, is granted by Humana Press Inc., provided that the base fee of US $10.00 per copy, plus US $00.25 per page, is paid directly to the Copyright Clearance Center at 222 Rosewood Drive, Danvers, MA 01923. For those organizations that have been granted a photocopy license from the CCC, a separate system of payment has been arranged and is acceptable to Humana Press Inc. The fee code for users of the Transactional Reporting Service is: [1-58829-291-6 $10.00 + $00.25].

Printed in the United States of America. 10 9 8 7 6 5 4 3 2 1

Library of Congress Cataloging in Publication Data
Functional genomics: methods and protocols / edited by Michael J. Brownstein and Arkady B. Khodursky.
      p. cm.—(Methods in molecular biology ; v. 224)
   Includes bibliographical references and index.
   ISBN 1-58829-291-6 (alk. paper) eISBN 1-59259-364-X
   1. DNA microarrays–Laboratory manuals. 2. Genomics–Laboratory manuals. 3. Proteomics–Laboratory manuals. I. Brownstein, Michael J. II. Khodursky, Arkady B. III. Methods in molecular biology (Totowa, NJ) ; v. 224
   QP624.5.D726F86 2003
      572.8'636–dc21
                                                                                              2003040707

Preface

Dramatic technological advances have marked every quantum leap in our understanding of biological systems. Advances in manipulating and sequencing DNA triggered the last such leap. More than 20 years ago this technology began finding its way into biology laboratories, and multiple landmark discoveries have followed, including the sequencing of the genomes of several prokaryotic and eukaryotic species. This, in turn, spawned a new interest in bioinformatics, structural biology, and high-throughput methods that would allow scientists to look at the response of the entire genome to physiological, pharmacological, and pathological changes. Whether this transition was driven by exhaustion of the existing hypothesis pool, a subconscious adoption of the old "new paradigm" of biological complexity, or an instinctual urge among biologists to look for new and interesting phenomena is not really important. What is important is that high-throughput methods are becoming more and more routine and available, and experimentalists and theoreticians must be prepared to take advantage of them. Spotted DNA microarrays fall into this category. They power functional genomics, a nascent research field dealing with the structure and activity of genomes and global relationships between genotype and phenotype. The birth of the field can be traced back to three seminal papers *(2,3,12)*. These three works grew from the realization that by placing individual sequence elements on a solid surface one can probe by hybridization a nearly unlimited number of targets simultaneously. Since then, similar *ad infinitum* approaches have been used to monitor relative protein levels *(6)*, protein functions *(15)*, cellular activities *(16)*, and molecular interactions *(10)*. These breakthrough studies will set the stage for the development of the fields of functional proteomics, metabolomics, etc. However, at the moment only functional genomic techniques on solid surfaces enjoy relatively wide acceptance because they are based on sounder physical-chemical principles.

The long-term impact of functional genomics will depend on multiple factors— standardization and simplification of the protocols used, robust error assessment, streamlining of the analytical techniques, the sustainability of cost. The goal of the volume is to familiarize its readers with available, reproducible protocols in the field, and to attempt to introduce an audience of biologists to data processing techniques that will become critically important as we start dealing with increasing quantities of information. The "Methods" are divided in two sections: (i) Methods in Data Generation and (ii) Methods in Data Analysis. The first section focuses on bench techniques that have been developed and are being routinely used in several hard-core genomics laboratories. Besides general applicability of the techniques, the articles represented in the first section were selected on the basis of one major criterion: that they give sufficiently robust protocols to be adopted without modification by workers who have just begun their journeys in the field of genomics. The section opens with articles describing ways to manufacture and use spotted microarrays on three different solid surfaces: glass, plastic, and nylon membranes. Arrays manufactured on glass surfaces are usually interrogated with fluorescent nucleic acid probes, and Chapter 4 describes an optimized RNA labeling procedure that is applicable to the known spectrum of RNA sources. This chapter is followed by articles dealing with issues and protocols that are common to the field of bacterial functional genomics. The last two years' work in the field of functional genomics was marked by development of specialized applications that have added to its depth. In this period there were techniques introduced that allow one to monitor subcellular RNA localization *in masse (4,9,14)*, to map chromosomes at the resolution of a single gene *(8,13)*, and to survey the steady-state genome-wide distribution of DNA binding proteins in vivo *(5,7,11)*. Chapters 6–8 deal primarily with the methodologies behind these advances; Chapter 7 also provides a link between expression profiling and determination of gene copy number using whole-genome DNA microarrays.

The issues of inference, experimental design, and reproducibility are of the paramount importance to researchers who deal with massive data sets. The second section of the "Methods" volume focuses on experimental design, data analysis, data display techniques, and bioinformatics. The section opens with a comprehensive overview of the inferential issues in microarray data analysis. The next four articles (Chapters 10–13) address in sequential manner some of the issues outlined in the overview article *(9)*: design of microarray experiments *(10)*; choice of the test statistic and assessment of the significance of observations *(11)*; data reduction *(12)*, and clustering, the most popular technique for microarray data classification *(13)*. Accelerating and making the most of functional genomics studies are impossible without visualization and data storage, both of which—like the remaining methods in the field—are still in their infancy. These problems cannot be overlooked, however, because they allow genomics researchers and their colleagues in the scientific community to examine and mine experimental results. Two articles *(14* and *15)* describe some available approaches to the data visualization and database related issues. There are several things that were not reflected in any of the featured articles but are nevertheless worth noting. Spotted DNA microarrays are currently being used mainly for two purposes: 1. screening and 2. modeling. Truly successful application of microarrays in these two areas depends, albeit to different degrees, on technology standardization as well as the free, unimpeded flow of experimental data. Although we believe that methods will become fairly standard in the near future, a good deal of useful and valuable information will be lost in the short run owing to the lack of enforceable standards and controlled vocabularies for experimental annotation. The Minimum Information About a Microarray Experiment (MIAME) specifically addresses this issue *(1)*, and should be read by anyone who wants to engage in expression profiling.

The "methods" we have compiled do not provide advice about or comparisons of the available spotting platforms, image extraction and analysis algorithms, data storage and retrieval devices, and data analysis compendia. Although the available options range from simple, relatively affordable solutions to high-end, sometimes extremely expensive, commercial ones, we believe that information accumulated in the field is not yet sufficient and/or systematic enough to provide comprehensive comparisons of individual solutions and/or specific recommendations.

In the course of preparing this volume we surveyed available microarray web resources. We encourage readers interested in developments in the field to keep an eye on the following sites:

> http://www.microarrays.org
> http://derisilab.ucsf.edu/
> http://www.mged.org/Workgroups/MIAME/miame.html
> http://www.bioconductor.org/
> http://ihome.cuhk.edu.hk/~b400559

Michael J. Brownstein
Arkady B. Khodursky

References

1. Brazma, A., P. Hingamp, Quackenbush, J., et al. (2001) Minimum information about a microarray experiment (MIAME)–toward standards for microarray data. *Nat. Genet.* **29,** 365–371.
2. Chuang, S. E., Daniels, D. L., and Blattner, F. R. (1993) Global regulation of gene expression in *Escherichia coli. J. Bacteriol.* **175,** 2026–2036.
3. DeRisi, J. L., Iyer, V. R., and Brown, P. O. (1997) Exploring the metabolic and genetic control of gene expression on a genomic scale. *Science* **278,** 680–686.
4. Diehn, M., Eisen, M. B., Botstein, D., and Brown, P. O. (2000) Large-scale identification of secreted and membrane-associated gene products using DNA microarrays. *Nat. Genet.* **25,** 58–62.
5. Gerton, J. L., DeRisi, J., Shroff, R., Lichten, M., Brown, P. O., and Petes, T. D. (2000) Inaugural article: global mapping of meiotic recombination hotspots and coldspots in the yeast *Saccharomyces cerevisiae. Proc. Natl. Acad. Sci. USA* **97,** 11383–11390.

6. Haab, B. B., Dunham, M. J., and Brown, P. O. (2001) Protein microarrays for highly parallel detection and quantitation of specific proteins and antibodies in complex solutions. *Genome Biol.* **2,** RESEARCH0004.
7. Iyer, V. R., Horak, C. E., Scafe, C. S., Botstein, D., Snyder, M., and Brown, P. O. (2001) Genomic binding sites of the yeast cell-cycle transcription factors SBF and MBF. *Nature* **409,** 533–538.
8. Khodursky, A. B., Peter, B. J., Schmid, M. B., et al. (2000) Analysis of topoisomerase function in bacterial replication fork movement: use of DNA microarrays. *Proc. Natl. Acad. Sci. USA* **97,** 9419–9424.
9. Kuhn, K. M., DeRisi, J. L., Brown, P. O., and Sarnow, P. (2001) Global and specific translational regulation in the genomic response of *Saccharomyces cerevisiae* to a rapid transfer from a fermentable to a nonfermentable carbon source. *Mol. Cell Biol.* **21,** 916–927.
10. Kuruvilla, F. G., Shamji, A. F., Sternson, S. M., Hergenrother, P. J., and Schreiber, S. L. (2002) Dissecting glucose signalling with diversity-oriented synthesis and small-molecule microarrays. *Nature* **416,** 653–657.
11. Lieb, J. D., Liu, X., Botstein, D., and Brown, P. O. (2001) Promoter-specific binding of Rap1 revealed by genome-wide maps of protein-DNA association. *Nat. Genet.* **28,** 327–334.
12. Nelson, S. F., McCusker, J. H., Sander, M. A., Kee, Y., Modrich, P., and Brown, P. O. (1993) Genomic mismatch scanning: a new approach to genetic linkage mapping. *Nat. Genet.* **4,** 11–18.
13. Pollack, J. R., Perou, C. M., Alizadeh, A. A., et al. (1999) Genome-wide analysis of DNA copy-number changes using cDNA microarrays. *Nat. Genet.* **23,** 41–46.
14. Takizawa, P. A., DeRisi, J. L., Wilhelm, J. E., and Vale, R. D. (2000) Plasma membrane compartmentalization in yeast by messenger RNA transport and a septin diffusion barrier. *Science* **290,** 341–344.
15. Zhu, H., Klemic, J. F., Chang, S., et al. (2000) Analysis of yeast protein kinases using protein chips. *Nat. Genet.* **26,** 283–289.
16. Ziauddin, J. and Sabatini, D. M. (2001) Microarrays of cells expressing defined cDNAs. *Nature* **411,** 107–110.

Contents

Preface v
Contributors xi

1　Fabrication of cDNA Microarrays
　　Charlie C. Xiang and Michael J. Brownstein 1

2　Nylon cDNA Expression Arrays
　　*George Jokhadze, Stephen Chen, Claire Granger,
　　and Alex Chenchik* 9

3　Plastic Microarrays: *A Novel Array Support Combining
　　the Benefits of Macro- and Microarrays*
　　*Alexander Munishkin, Konrad Faulstich,
　　Vissarion Aivazachvili, Claire Granger, and Alex Chenchik* 31

4　Preparing Fluorescent Probes for Microarray Studies
　　Charlie C. Xiang and Michael J. Brownstein 55

5　*Escherichia coli* Spotted Double-Strand DNA Microarrays:
　　*RNA Extraction, Labeling, Hybridization, Quality Control,
　　and Data Management*
　　*Arkady B. Khodursky, Jonathan A. Bernstein, Brian J. Peter,
　　Virgil Rhodius, Volker F. Wendisch, and Daniel P. Zimmer* 61

6　Isolation of Polysomal RNA for Microarray Analysis
　　Yoav Arava 79

7　Parallel Analysis of Gene Copy Number and Expression Using
　　cDNA Microarrays
　　Jonathan R. Pollack 89

8　Genome-wide Mapping of Protein–DNA Interactions by Chromatin
　　Immunoprecipitation and DNA Microarray Hybridization
　　Jason D. Lieb 99

9　Statistical Issues in cDNA Microarray Data Analysis
　　Gordon K. Smyth, Yee Hwa Yang, and Terry Speed 111

10　Experimental Design to Make the Most of Microarray Studies
　　M. Kathleen Kerr 137

11 Statistical Methods for Identifying Differentially Expressed
   Genes in DNA Microarrays
   John D. Storey and Robert Tibshirani 149

12 Detecting Stable Clusters Using Principal Component Analysis
   Asa Ben-Hur and Isabelle Guyon 159

13 Clustering in Life Sciences
   Ying Zhao and George Karypis 183

14 A Primer on the Visualization of Microarray Data
   Paul Fawcett 219

15 Microarray Databases: *Storage and Retrieval of Microarray Data*
   Gavin Sherlock and Catherine A. Ball 235

Index 249

Contributors

VISSARION AIVAZACHVILI • *BD Biosciences Clontech, Palo Alto, CA*

YOAV ARAVA • *Department of Biochemistry, Stanford University School of Medicine, Palo Alto, CA*

CATHERINE A. BALL • *Department of Genetics, Stanford University School of Medicine, Palo Alto, CA*

ASA BEN-HUR • *Department of Biochemistry, Stanford University School of Medicine, Palo Alto, CA*

JONATHAN A. BERNSTEIN • *Department of Genetics, Stanford University School of Medicine, Palo Alto, CA*

MICHAEL J. BROWNSTEIN • *Laboratory of Genetics, NIMH/NHGRI, National Institutes of Health, Rockville, MD*

STEPHEN CHEN • *BD Biosciences Clontech, Palo Alto, CA*

ALEX CHENCHIK • *BD Biosciences Clontech, Palo Alto, CA*

KONRAD FAULSTICH • *BD Biosciences Clontech, Palo Alto, CA*

PAUL FAWCETT • *Department of Biochemistry, Stanford University School of Medicine, Palo, CA*

CLAIRE GRANGER • *BD Biosciences Clontech, Palo Alto, CA*

ISABELLE GUYON • *Clopinet, Berkeley, CA*

GEORGE JOKHADZE • *BD Biosciences Clontech, Palo Alto, CA*

GEORGE KARYPIS • *Department of Computer Science, University of Minnesota, Minneapolis, MN*

M. KATHLEEN KERR • *Department of Biostatistics, University of Washington, Seattle, WA*

ARKADY B. KHODURSKY • *Department of Biochemistry, Biophysics and Molecular Biology, University of Minnesota, St. Paul, MN*

JASON D. LIEB • *Department of Biology and Carolina Center for Genome Sciences, The University of North Carolina at Chapel Hill, Chapel Hill, NC*

ALEXANDER MUNISHKIN • *BD Biosciences Clontech, Palo Alto, CA*

BRIAN J. PETER • *MRC Laboratory of Molecular Biology, Cambridge, UK*

JONATHAN R. POLLACK • *Department of Pathology, Stanford University School of Medicine, Palo Alto, CA*

VIRGIL RHODIUS • *Department of Stomatology, University of California, San Francisco, San Francisco, CA*

GAVIN SHERLOCK • *Department of Genetics, Stanford University School of Medicine, Palo Alto, CA*

GORDON K. SMYTH • *Walter and Eliza Hall Institute of Medical Research, Melbourne, Australia*

TERRY SPEED • *Walter and Eliza Hall Institute of Medical Research, Melbourne, Australia; and Department of Statistics, University of California–Berkeley, Berkeley, CA*

JOHN D. STOREY • *Department of Statistics, Stanford University, Palo Alto, CA*

ROBERT TIBSHIRANI • *Department of Health Research & Policy and Department of Statistics, Stanford University, Palo Alto, CA*

VOLKER F. WENDISCH • *Institute of Biotechnology, Juelich, Germany*

CHARLIE C. XIANG • *Laboratory of Genetics, NIMH/NHGRI, National Institutes of Health, Rockville, MD*

YEE HWA YANG • *Division of Biostatistics, University of California, San Francisco, CA*

YING ZHAO • *Department of Computer Science, University of Minnesota, Minneapolis, MN*

DANIEL P. ZIMMER • *Microbia Inc., Cambridge, MA*

1

Fabrication of cDNA Microarrays

Charlie C. Xiang and Michael J. Brownstein

1. Introduction

DNA microarray technology has been used successfully to detect the expression of many thousands of genes, to detect DNA polymorphisms, and to map genomic DNA clones *(1–4)*. It permits quantitative analysis of RNAs transcribed from both known and unknown genes and allows one to compare gene expression patterns in normal and pathological cells and tissues *(5,6)*.

DNA microarrays are created using a robot to spot cDNA or oligonucleotide samples on a solid substrate, usually a glass microscope slide, at high densities. The sizes of spots printed in different laboratories range from 75 to 150 µm in diameter. The spacing between spots on an array is usually 100–200 µm. Microarrays with as many as 50,000 spots can be easily fabricated on standard 25 mm × 75 mm glass microscope slides.

Two types of spotted DNA microarrays are in common use: cDNA and synthetic oligonucleotide arrays *(7,8)*. The surface onto which the DNA is spotted is critically important. The ideal surface immobilizes the target DNAs, and is compatible with stringent probe hybridization and wash conditions *(9)*. Glass has many advantages as such a support. DNA can be covalently attached to treated glass surfaces, and glass is durable enough to tolerate exposure to elevated temperatures and high-ionic-strength solutions. In addition, it is nonporous, so hybridization volumes can be kept to a minimum, enhancing the kinetics of annealing probes to targets. Finally, glass allows probes labeled with two or more fluors to be used, unlike nylon membranes, which are typically probed with one radiolabeled probe at a time.

2. Materials

1. Multiscreen filtration plates (Millipore, Bedford, MA).
2. Qiagen QIAprep 96 Turbo Miniprep kit (Qiagen, Valencia, CA).
3. dATP, dGTP, dCTP, and dTTP (Amersham Pharmacia, Piscataway, NJ).
4. M13F and M13R primers (Operon, Alameda, CA).
5. *Taq* DNA polymerase and buffer (Invitrogen, Carlsbad, CA).
6. PCR CyclePlate (Robbins, Sunnyvale, CA).
7. CycleSeal polymerase chain reaction (PCR) plate sealer (Robbins).
8. Gold Seal microscope slides (Becton Dickinson, Franklin, NJ).
9. 384-well plates (Genetix, Boston, MA).
10. Succinic anhydride (Sigma, St. Louis, MO) in 325 mL of 1-methy-2-pyrrolidinone (Sigma).

3. Methods
3.1. Selection and Preparation of cDNA Clones
3.1.1. Selection of Clones

Microarrays are usually made with DNA fragments that have been amplified by PCR from plasmid samples or directly from chromosomal DNA. The sizes of the PCR products on our arrays range from 0.5 to 2 kb. They attach well to the glass surface. The amount of DNA deposited per spot depends on the pins chosen for printing, but elements with 250 pg to 1 ng of DNA (up to 9×10^8 molecules) give ample signals.

Many of the cDNA clones that have been arrayed by laboratories in the public domain have come from the Integrated Molecular Analysis of Genomes and Expression (IMAGE) Consortium set. Five million human IMAGE clones have been collected and are available from Invitrogen/Research Genetics (www.resgen.com/products/IMAGEClones.php3). Sequence-verified cDNA clones from humans, mice, and rats are also available from Invitrogen/Research Genetics.

cDNA clones can also be obtained from other sources. The 15,000 National Institute of Aging (NIA) mouse cDNA set has been distributed to many academic centers (http://lgsun.grc.nia.nih.gov/cDNA/15k/hsc.html). Other mouse cDNA collections include the Brain Molecular Anatomy Project (BMAP) (http://brainest.eng.uiowa.edu), and RIKEN (http://genome.rtc.riken.go.jp) clone sets. In preparing our arrays, we have used the NIA and BMAP collections and are in the process of sequencing the 5′ ends of the 41,000 clones in the combined set in collaboration with scientists at the Korea Research Institute of Bioscience and Biotechnology. Note that most cDNA collections suffer from some gridding errors and well-to-well cross contamination.

3.1.2. Preparation of Clones

Preparing DNA for spotting involves making plasmid minipreps, amplifying their inserts, and cleaning up the PCR products. Most IMAGE clones are in standard cloning vectors, and the inserts can be amplified with modified M13 primers. The sequences of the forward (M13F) and reverse (M13R) primers used are 5′-GTTGTAAAACGACGGCCAGTG-3′ and 5′-CACACAGGAAA CAGCTATG-3′, respectively. A variety of methods are available for purifying cDNA samples. We use QIAprep 96 Turbo Miniprep kits and a Qiagen BioRobot 8000 (Qiagen) for plasmid isolations but cheaper, semiautomated techniques can be used as well. We PCR DNAs with a Tetrad MultiCycler (MJ Research, Incline Village, NV) and purify the products with Multiscreen filtration plates (Millipore).

3.1.3. Purification of Plasmid

1. Culture the bacterial clones overnight in 1.3 mL of Luria–Bertani (LB) medium containing 100 µg/mL of carbenicillin at 37°C, shaking them at 300 rpm in 96-well flat-bottomed blocks.
2. Harvest the bacteria by centrifuging the blocks for 5 min at 1500g in an Eppendorf centrifuge 5810R (Eppendorf, Westbury, NY). Remove the LB by inverting the block. The cell pellets can be stored at –20°C.
3. Prepare cDNA using the BioRobot 8000, or follow the Qiagen QIAprep 96 Turbo Miniprep kit protocol for manual extraction.
4. Elute the DNA with 100 µL of Buffer EB (10 mM Tris-HCl, pH 8.5) included in the QIAprep 96 Turbo Miniprep kit. The plasmid DNA yield should be 5–10 µg per prep.

3.1.4. PCR Amplification

1. Dilute the plasmid solution 1:10 with 1X TE (10 mM Tris-HCl, pH 8.0, 1 mM EDTA).
2. For each 96-well plate to be amplified, prepare a PCR reaction mixture containing the following ingredients: 1000 µL of 10X PCR buffer (Invitrogen), 20 µL each of dATP, dGTP, dCTP, and dTTP (100 mM each; Amersham Pharmacia), 5 µL each of M13F and M13R (1 mM each; Operon), 100 µL of Taq DNA polymerase (5 U/µL; Invitrogen), and 8800 µL of ddH$_2$O.
3. Add 100 µL of PCR reaction mix to each well of a PCR CyclePlate (Robbins) plus 5 µL of diluted plasmid template. Seal the wells with CycleSeal PCR plate sealer (Robbins). (Prepare two plates for amplification from each original source plate to give a final volume of 200 µL of each product.)
4. Use the following PCR conditions: 96°C for 2 min; 30 cycles at 94°C for 30 s, 55°C for 30 s, 72°C for 1 min 30 s; 72°C for 5 min; and cool to ambient temperature.

5. Analyze 2 µL of each product on 2% agarose gels. We use an Owl Millipede A6 gel system (Portsmouth, NH) with eight 50-tooth combs. This allows us to run 384 samples per gel.

3.1.5. Cleanup of PCR Product

1. Transfer the PCR products from the two duplicate PCR CyclePates to one Millipore Multiscreen PCR plate using the Qiagen BioRobot 8000.
2. Place the Multiscreen plate on a vacuum manifold. Apply the vacuum to dry the plate.
3. Add 100 µL of ddH_2O to each well.
4. Shake the plate for 30 min at 300 rpm.
5. Transfer the purified PCR products to a 96-well plate.
6. Store the PCR products in a –20°C freezer.

3.2. Creating cDNA Microarrays (see Note 1)

Robots are routinely used to apply DNA samples to glass microscope slides. The slides are treated with poly-L-lysine or other chemical coatings. Some investigators irradiate the printed arrays with UV light. Slides coated with poly-L-lysine have a positively charged surface, however, and the negatively charged DNA molecules bind quite tightly without crosslinking. Finally, the hydrophobic character of the glass surface minimizes spreading of the printed spots. Poly-L-lysine-coated slides are inexpensive to make, and we have found that they work quite well.

About 1 nL of PCR product is spotted per element. Many printers are commercially available. Alternatively, one can be built in-house (for detailed instructions, visit http://cmgm.stanford.edu/pbrown/mguide/index.html). After the arrays are printed, residual amines are blocked with succinic anhydride (see http://cmgm.stanford.edu/pbrown/mguide/index.html).

3.2.1. Coating Slides with Poly-L-lysine

1. Prepare cleaning solution by dissolving 100 g of NaOH in 400 mL of ddH_2O. Add 600 mL of absolute ethanol and stir until the solution clears.
2. Place Gold Seal microscope slides (Becton Dickinson) into 30 stainless-steel slide racks (Wheaton, Millville, NJ). Place the racks in a glass tank with 500 mL of cleaning solution. Work with four racks (120 slides in total) at a time.
3. Shake at 60 rpm for 2 h.
4. Wash with ddH_2O four times, 3 min for each wash.
5. Make a poly-L-lysine solution by mixing 80 mL of 0.1% (w/v) poly-L-lysine with 80 mL of phosphate-buffered saline and 640 mL of ddH_2O.
6. Transfer two racks into one plastic tray with 400 mL of coating solution.
7. Shake at 60 rpm for 1 h.

8. Rinse the slides three times with ddH$_2$O.
9. Dry the slides by placing them in racks (Shandon Lipshaw, Pittsburgh, PA) and spinning them at 130g for 5 min in a Sorvall Super T21 centrifuge with an ST-H750 swinging bucket rotor. Place one slide rack in each bucket.
10. Store the slides in plastic storage boxes and age them for 2 wk before printing DNA on them.

3.2.2. Spotting DNA on Coated Slides

We use the following parameters to print 11,136 element arrays with an OmniGrid robot having a Server Arm (GeneMachines, San Carlos, CA): 4 × 4 SMP3 pins (TeleChem, Sunnyvale, CA), 160 × 160 µM spacing, 27 × 26 spots in each subarray, single dot per sample. We use the following printing parameters: velocity of 13.75 cm/s, acceleration of 20 cm/s^2, deceleration of 20 cm/s^2. We print two identical arrays on each slide.

1. Adjust the relative humidity of the arrayer chamber to 45–55% and the temperature to 22°C.
2. Dilute the purified PCR products 1:1 with dimethylsulfoxide (DMSO) (Sigma) (*see* **Note 2**). Transfer 10-µL aliquots of the samples to Genetix 384-well plates (Genetix).
3. Load the plates into the cassette of the Server Arm. Three such cassettes hold 36 plates. Reload the cassettes in midrun if more than 36 plates of samples are to be printed. It takes about 24 h to print 100 slides with 2 × 11,136 elements on them.
4. Label the slides. Examine the first slide in the series under a microscope. Mark the four corners of the array (or the separate arrays if there are more than one on the slide) with a scribe. Use this indexed slide to draw a template on a second microscope slide showing where the cover slip should be placed during the hybridization step. Remove the remaining slides from the arrayer and store them in a plastic box.

3.2.3. Postprocessing

We often postprocess our arrays after storing them for several days. This may not be necessary as others have argued, but it is sometimes convenient. Many workers recommend UV crosslinking the DNA to the slide surface by exposing the arrays to 450 mJ of UV irradiation in a Stratalinker (Stratagene, La Jolla, CA). As noted, this step is optional, and we have not found it to be critical.

1. Insert 30 slides into a stainless steel rack and place each rack in a small glass tank.
2. In a chemical fume hood, dissolve 6 g of succinic anhydride (Sigma) in 325 mL of 1-methy-2-pyrrolidinone (Sigma) in a glass beaker by stirring.

3. Add 25 mL of 1 M sodium borate buffer (pH 8.0) to the beaker as soon as the succinic anhydride is dissolved.
4. Rapidly pour the solution into the glass tank.
5. Place the glass tank on a platform shaker and shake at 60 rpm for 20 min in the hood. While the slides are incubating on the shaker, prepare a boiling water bath.
6. Transfer the slides to a container with 0.1% sodium dodecyl sulfate solution. Shake at 60 rpm for 3 min.
7. Wash the slides with ddH$_2$O for 2 min. Repeat the wash two more times.
8. Place the slides in the boiling water bath. Turn off the heat immediately after submerging the slides in the water. Denature the DNA for 2 min in the water bath.
9. Transfer the slides to a container with 100% ethanol and incubate for 4 min.
10. Dry the slides in a centrifuge at 130g for 5 min (*see* **Subheading 3.2.1., step 9**) and store them in a clean plastic box. The slides are now ready to be probed (*see* **Note 3**).

4. Notes

1. The methods for printing slides described in this chapter are somewhat tedious, but they are robust and inexpensive.
2. We recommend dissolving the DNAs to be printed in 50% DMSO instead of aqueous buffers because this is a simple solution to the problem of sample evaporation during long printing runs *(10)*.
3. The probe-labeling technique that we describe in Chapter 4 works well with slides prepared according to the protocols we have given.

References

1. Schena, M., Shalon, D., Davis, R. W., and Brown, P. O. (1995) Quantitative monitoring of gene expression patterns with a complementary DNA microarray. *Science* **270,** 467–470.
2. Schena, M., Shalon, D., Heller, R., Chai, A., Brown, P. O., and Davis, R. W. (1996) Parallel human genome analysis: microarray-based expression monitoring of 1000 genes. *Proc. Natl. Acad. Sci. USA* **93,** 10,614–10,619.
3. DeRisi, J., Vishwanath, R. L., and Brown, P. O. (1997) Exploring the metabolic and genetic control of gene expression on a genomic scale. *Science* **278,** 680–686.
4. Sapolsky, R. J. and Lipshutz, R. J. (1996) Mapping genomic library clones using oligonucleotide arrays. *Genomics* **33,** 445–456.
5. DeRisi, J., Penland, L., Brown, P. O., Bittner, M. L., Meltzer, P. S., Ray, M., Chen, Y., Su, Y. A., and Trent, J. M. (1996) Use of a cDNA microarray to analyse gene expression patterns in human cancer. *Nat. Genet.* **14,** 457–460.
6. Heller, R. A., Schena, M., Chai, A., Shalon, D., Bedilion, T., Gilmore, J., Woolley, D. E., and Davis, R. W. (1997) Discovery and analysis of inflammatory disease-related genes using cDNA microarray. *Proc. Natl. Acad. Sci. USA* **94,** 2150–2155.

7. Shalon, D., Smith, S. J., and Brown, P. O. (1996) A DNA microarray system for analyzing complex DNA samples using two-color fluorescent probe hybridization. *Genome Res.* **6,** 639–645.
8. Lipshutz, R. J., Fodor, S. P. A., Gingeras, T. R., and Lockhart, D. J. (1999). High density synthetic oligonucleotide arrays. *Nat. Genet.* **21(Suppl.),** 20–24.
9. Cheung, V. G., Morley, M., Aguilar, F., Massimi, A., Kucherlapati, R., and Childs, G. (1999) Making and reading microarrays. *Nat. Genet.* **21(Suppl.),** 15–19.
10. Hegde, P., Qi, R., Abernathy, K., Gay, C., Dharap, S., Gaspard, R., Hughes, J. E., Snesrud, E., Lee, N., and Quackenbush, J. (2000) A concise guide to cDNA microarray analysis. *Biotechniques* **29,** 548–556.

2

Nylon cDNA Expression Arrays

George Jokhadze, Stephen Chen, Claire Granger, and Alex Chenchik

1. Introduction

Nucleic acid arrays provide a powerful methodology for studying biological systems on a genomic scale. BD Atlas™ Arrays, developed by BD Biosciences Clontech, are expression profiling products specifically designed to be accessible to all laboratories performing isotopic blot hybridization experiments. We have developed two types of readily accessible BD Atlas Arrays: nylon macroarrays, well suited for high-sensitivity expression profiling using a limited gene set, and broad-coverage plastic microarrays, for a more extensive analysis of a comprehensive set of genes. In this chapter, we describe protocols for printing and performing gene expression analysis using nylon membrane–based arrays. For a more in-depth description and protocols related to plastic film–based arrays, please refer to Chapter 3.

Nylon membrane–based arrays offer several advantages for researchers. Compared with glass arrays, nylon arrays are usually less expensive to produce and require less complicated equipment. Nylon arrays are generally considered more user friendly, since analysis involves only familiar hybridization techniques. Detection of results is also straightforward—probes are radioactively labeled, so one can simply use a standard phosphorimager.

1.1. Sensitivity of Nylon Arrays

Nylon membranes are typically used to print low- (10–1000) to medium- (1000–4000) density cDNA arrays. Unlike high-density arrays, which are usually printed on glass or plastic supports, probes for nylon arrays can be labeled with ^{32}P, resulting in a much higher (>fourfold) level of sensitivity

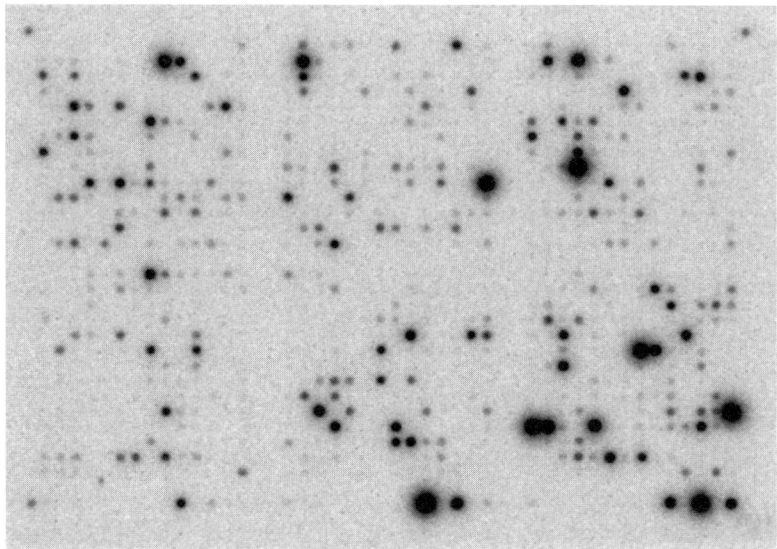

Fig. 1. Nylon array hybridized with a ^{32}P-labeled probe.

(**Fig. 1**). This means that the presence of even low-abundance transcripts can be detected.

Nylon arrays are printed with fragments of cDNA clones (200–600 bp) representing individual genes. Each cDNA fragment is amplified from the original clone using gene-specific or universal primers, denatured, and printed onto the membranes. cDNA fragments have a significantly higher hybridization efficiency than oligos yet generally do not allow discrimination between highly homologous genes, such as multigene family members. For this reason, cDNA fragments are ideal for nylon arrays that represent a limited number of genes. In an array experiment, the cDNA fragments on the array are designated as the "targets." The "probe" used to screen the array is a radioactively labeled pool of cDNAs, reverse transcribed from total or polyA$^+$ RNA extracted from a particular tissue or cell type. Duplicate arrays are screened with cDNA probes prepared from two or more tissues, cell lines, or differentially treated samples.

The single most important factor determining the success or failure of array experiments is the quality of the RNA used to make the probes. Poor-quality RNA preparation leads to high background on the membrane and/or a misleading hybridization pattern. The present protocol allows purification of total RNA and labeling of probes for array hybridizations in one straightforward procedure—no separate poly A$^+$ RNA purification step is needed. An acceptable amount (10 µg) of high-quality total RNA can be isolated from as little as 10 mg of tissue or 10^5 cells.

With nylon membrane arrays, there is a choice of using ^{32}P or ^{33}P in the labeling reaction. The more appropriate method depends on the printing density of the array (see **Subheading 3.1.4.**) and the nature of the experiment. For general purposes, we recommend using ^{32}P because this isotope provides greater sensitivity. High sensitivity will be especially important if one is interested in any low-abundance transcripts. On the other hand, ^{33}P offers the advantage of higher-resolution signal, meaning that the signal produced by a spot on the array will be more closely confined to the spot's center, preventing signal "bleed" to neighboring spots. High signal bleed can complicate the interpretation of results for nearby genes. The ^{33}P method is particularly useful if highly abundant transcripts are of interest or one plans to quantitatively analyze the results by phosphorimaging. However, ^{33}P detection is generally only one-fourth as sensitive as ^{32}P detection *(1)*. When labeling array probes, choose the method that best suits your needs.

2. Materials

Unless otherwise noted, all catalog numbers provided are for BD Biosciences Clontech products.

2.1. Nylon Membrane Array Printing

2.1.1. Nylon Membrane Printing Reagents

1. Nytran Plus Membrane, cut into 82 × 120 mm rectangles (Schleicher & Schuell).
2. BD TITANIUM™ *Taq* PCR Kit (cat. no. K1915-1).
3. Gene-specific or universal primers for amplifying cDNA fragments (*see* **Subheading 3.1.**).
4. Sequence-verified cDNA templates (vectors carrying clones with sequence-verified cDNA insert).
5. Milli-Q-filtered H_2O.
6. Printing dye (30% Ficoll, 1% thymol blue).
7. 3 *M* NaOAc, pH 4.0.
8. Membrane neutralization solution (0.5 *M* Tris, pH 7.6).

2.1.2. Nylon Membrane Array Printing Equipment

1. Polymerase chain reaction (PCR) reaction tubes (0.5 mL). (We recommend Perkin-Elmer GeneAmp 0.5-mL reaction tubes (cat. no. N801-0737 or N801-0180).
2. PCR machine/thermal cycler. We use a hot-lid thermal cycler.

3. 384-well V-bottomed polystyrene plates (USA Scientific), for use as a source plate during printing.
4. SpeedVac.
5. Arrayer robot. We use a *Bio*Grid Robot (*Bio*Robotics).
6. UV Stratalinker crosslinker (Stratagene).
7. Pin tool (0.7 mm diameter, 384 pin).
8. Sarstedt Multiple Well Plate 96-Well (lids only), used to hold nylon membranes for printing.
9. Adhesive sealing film (THR100 Midwest Scientific).
10. NucleoSpin Multi-8 PCR Kit (cat. no. K3059-1) or NucleoSpin Multi-96 PCR Kit (cat. no. K3065-1).

2.2. Reagents for RNA Isolation and Probe Synthesis

2.2.1. Reagents Provided with BD Atlas Pure Total RNA Labeling System

The BD Atlas™ Pure Total RNA Labeling System (cat. no. K1038-1) is available exclusively from BD Biosciences Clontech. Do not use the protocol supplied with the BD Atlas Pure Kit. The procedures for RNA isolation and cDNA synthesis in the following protocol differ significantly from the procedures found in the BD Atlas Pure User Manual.

1. Denaturing solution.
2. Saturation buffer for phenol.
3. RNase-free H_2O.
4. 2 M NaOAc (pH 4.5).
5. 10X termination mix.
6. Streptavidin magnetic beads.
7. 1X binding buffer.
8. 2X binding buffer.
9. 1X reaction buffer.
10. 1X wash buffer
11. DNase I (1 U/µL).
12. DNase I buffer.
13. Biotinylated oligo(dT).
14. Moloney murine leukemia virus reverse transcriptase (MMLV RT).

2.2.2. Additional Reagents/Special Equipment

1. Saturated phenol (store at 4°C). For 160 mL: 100 g of phenol (Sigma cat. no. P1037 or Boehringer Mannheim cat. no. 100728). In a fume hood, heat a jar of phenol in a 70°C water bath for 30 min or until the phenol is completely melted. Add 95 mL of phenol directly to the saturation buffer (from the BD AtlasPure Kit), and mix well. Hydroxyquinoline may be added if desired. Aliquot and freeze at –20°C for long-term storage. This preparation of saturated phenol will only have one phase.

2. Tissue homogenizer (e.g., Polytron or equivalent). For <200 mg of tissue, use a 6-mm probe. For >200 mg of tissue, use a 10-mm probe.
3. [α-^{32}P]dATP (10 µCi/µL; 3000 Ci/mmol) (cat. no. PB10204; Amersham) or [α-^{33}P]dATP (10 µCi/µL; >2500 Ci/mmol) (cat. no. BF1001; Amersham). Do not use Amersham's Redivue or any other dye-containing isotope.
4. Deionized H_2O (Milli-Q filtered or equivalent; do not use diethylpyrocarbonate-treated H_2O).
5. Magnetic particle separator (cat. no. Z5331; Promega, Madison, WI). It is important that you use a separator designed for 0.5-mL tubes.
6. Polypropylene centrifuge tubes: 1.5-mL (cat. no. 72-690-051; Sarstedt), 2-mL (cat. no. 16-8105-75; PGC), 15-mL (tubes cat. no. 05-562-10D, caps cat. no. 05-562-11E; Fisher), and 50-mL (tubes with caps cat. no. 05-529-1D; Fisher). Fifteen- and 50-mL tubes should be sterilized with 1% sodium dodecyl sulfate (SDS) and ethanol before use.
7. 10X dNTP mix (for dATP label; 5 mM each of dCTP, dGTP, dTTP).
8. 10X Random primer mix (N-15) or gene-specific primer mix (*see* **Subheading 3.4.3.**).
9. BD PowerScript™ Reverse Transcriptase and 5X BD PowerScript™ Reaction Buffer (available exclusively from BD Biosciences Clontech; cat. no. 8460-1).
10. Dithiothreitol (DTT) (100 mM).
11. NucleoSpin® Extraction Kit: NucleoSpin extraction spin columns, 2-mL collection tubes, buffer NT2, buffer NT3 (add 95% ethanol before use as specified on the label), buffer NE.

2.3. Reagents for Hybridization, Washing, and Stripping of Nylon Arrays

1. BD ExpressHyb™ hybridization solution (cat. no. 8015-1).
2. Sheared salmon testes DNA (10 mg/mL) (cat. no. D7656; Sigma).
3. **Optional:** 10X Denaturing solution (1 M NaOH, 10 mM EDTA) (*see* **Subheading 3.5.**).
4. **Optional:** 2X Neutralizing solution (1 M NaH$_2$PO$_4$ [pH 7.0]): 27.6 g of NaH$_2$PO$_4$•H$_2$O). Add 190 mL of H_2O, adjust the pH to 7.0 with 10 N NaOH if necessary, and add H_2O to 200 mL. Store at room temperature (*see* **Subheading 3.5.**).
5. C$_0$t-1 DNA (1 mg/mL).
6. 20X saline sodium citrate (SSC), 175.3 g of NaCl, 88.2 g of Na$_3$citrate•2H$_2$O. Add 900 mL of H_2O, adjust the pH to 7.0 with 1 M HCl if necessary, and add H_2O to 1 L. Store at room temperature.
7. 20% SDS: 200 g of SDS. Add H_2O to 1 L. Heat to 65°C to dissolve. Store at room temperature.
8. Wash solution 1: 2X SSC, 1% SDS. Store at room temperature.
9. Wash solution 2: 0.1X SSC, 0.5% SDS. Store at room temperature.

3. Methods
3.1. Printing of Nylon Membrane Arrays

cDNA fragments to be used for printing can be amplified by using either gene-specific primers or a pair of "universal" primers (i.e., T3, T7, M13F, or M13R) complementary to sites in the cloning vector flanking the cDNA clone. One advantage of using gene-specific primers is that a specific region of the cDNA clone to be amplified can be chosen. For example, the amplification of cDNAs used to print BD Atlas Arrays is specially designed to minimize nonspecific hybridization. All cDNA fragments are 200–600 bp long and are amplified from a region of the mRNA that lacks the poly A tail, repetitive elements, or other highly homologous sequences. Another advantage of using gene-specific primers is that the antisense primers used in array preparation can be pooled and subsequently used as a gene-specific primer mix to synthesize cDNA probes from experimental samples. The use of gene-specific probes provides higher sensitivity and lower background than random primers (*see* **Subheading 3.4.3.** for details).

3.1.1. Preparative PCR for cDNA Fragments

1. Prepare a 100-µL PCR reaction in a 0.5-mL PCR tube for each cDNA to be represented on the array. Calculate the amount of each component required for the PCR reaction by referring to **Table 1**. Universal primers, appropriate for your cloning vector, may be used in place of gene-specific primers. Adjust the volumes accordingly.
2. Commence thermal cycling using the following parameters: 30–35 cycles of 94°C for 30 s and 68°C for 90 s, 68°C for 5 min, and 15°C soak. These conditions were developed for use with a hot-lid thermal cycler; the optimal parameters may vary with different thermal cyclers. (Note that these parameters were optimized for amplification of fragments approx 200–600 bp long.)
3. Run 5 µL of each pooled PCR product (plus loading dye) on a 2% TAE agarose gel, alongside a molecular weight marker, to screen the PCR products.
4. Check each PCR product size by comparison with the molecular weight markers. If the size of the PCR product is correct, add EDTA (final concentration of 0.1 M EDTA, pH 8.0) to the pooled PCR products to stop the reaction.

3.1.2. Purification of cDNA Fragments

To purify amplified cDNA fragments, we recommend that you use either the NucleoSpin Multi-8 PCR Kit (cat. no. K3059-1) or NucleoSpin Multi-96 PCR Kit (cat. no. K3065-1) and follow the enclosed protocol. NucleoSpin PCR kits are designed to purify PCR products from reaction mixtures with speed and efficiency. Primers, nucleotides, salts, and polymerases are effectively removed using these kits; up to 96 samples can be processed simultaneously in less than

Table 1
cDNA Fragment PCR Set-Up

PCR master mix	Final concentration	Per 100-µL reaction (µL)
10X BD TITANIUM *Taq* PCR buffer	1X	10
10 µM dNTP mix	200 µM	2
Specific or universal primer mix, 20 µM each	0.4 µM each	2
Template (0.5–1 ng/µL)	0.025–0.05 ng/µL	5
50X BD TITANIUM *Taq* Mix	1X	2
Milli-Q H$_2$O	Bring volume up	79

60 min. Up to 15 µg of high-quality DNA can be isolated per preparation. Recovery rates of 75–90% can be achieved for fragments from 100 bp to 10 kb.

3.1.3. Standardization of cDNAs

1. In a 1.5-mL microcentrifuge tube, dilute 5 µL of the purified cDNA fragment stock in 995 µL of H$_2$O (a 1:200 dilution) and read the optical density of the dilution at 260 nm. Calculate the cDNA concentration in cDNA stock. Each PCR reaction should yield a total of 2 to 3 µg of DNA.
2. If the concentration of cDNA in the stock solution is >500 ng/µL, go to **step 5**; if <500 ng/µL, continue with the next step.
3. Concentrate the cDNA stock solution by evaporation in a SpeedVac. Repeat **steps 1** and **2**.
4. Adjust the concentration to 500 ng/µL by adding Milli-Q-H$_2$O: $V_{H_2O} = (C_i \times V_i / C_f) - V_i$, in which C_i and V_i are the initial concentration and volume of the main solution (before adding H$_2$O), respectively; and C_f is the final, desired concentration.
5. Store the normalized cDNA at –20°C.

3.1.4. Printing of cDNA Arrays on Nylon Membranes

An 80 mm × 120 mm rectangle of nylon membrane can be printed with as many as 3000 cDNA fragments (using a 384-pin tool with 0.7-mm-diameter pins) without encountering significant difficulties with image analysis due to signal bleed. If ^{32}P-labeled probes are used, the maximum printing density on a membrane of the same size should be no more than 1500, to avoid loss of signal resolution.

Depending on your experimental needs and organism, you may wish to include negative controls, such as genomic DNA, phage lambda DNA, or yeast

DNA. Some researchers also choose to include cDNA fragments representing certain housekeeping genes, known to be highly expressed in their experimental samples, to serve as positive controls.

1. Prepare the individual cDNA printing mixes. The final cDNA concentration for printing should be approx 100 ng/µL. The final NaOH concentration for printing should be 0.15 N. The final printing dye concentration for printing should be 1X. The volume of solution deposited by a single, 0.7-mm-diameter pin is 90 nL, which is equivalent to 10 ng of cDNA printed per spot. For example, to prepare 25 µL of ready-to-print cDNA solution with a ~110 ng/µL final concentration, combine: 5.5 µL of cDNA (500 ng/µL), 0.4 µL of 10 N NaOH, 2.5 µL of 10X dye, and 16.6 µL of Milli-Q H_2O, for a total of 25.0 µL. This volume is sufficient for printing approx 200 arrays with single spots for each cDNA, or 100 arrays with duplicate spots. (Printing from volumes of <2 to 3 µL may result in irregular spot morphology.)
2. Aliquot 25 µL of each cDNA printing mix into individual wells of a 384-well plate.
3. Prepare the arrayer for printing following the manufacturer's user manual. (We use a *Bio*Robotics *Bio*Grid.)
4. Place each nylon membrane onto a lid from a Sarstedt 96-well plate. This will hold the membrane securely during printing. Place the Nytran Plus membranes and lids into the filter tray (the *Bio*grid tray holds 24 membranes at a time).
5. Begin the printing process according to the manufacturer's instructions.
6. Replace the water and ethanol in the arrayer's trays after every second round of printing.
7. After the completion of printing, allow the membranes to dry for 45 min at room temperature.
8. Using forceps, pick up the dried, printed membranes, grasping each membrane only by the edge, and drop into a tray containing membrane-neutralizing solution. Gently agitate the membrane arrays for approx 1 min. Change the solution after every 48 membranes.
9. Crosslink the membranes using an energy of 120 mJ/cm^2 (1200 × 100 µJ/cm^2) in a UV Stratalinker Crosslinker. When complete, remove the membranes from the Stratalinker and lay flat to dry for at least 4 h. Dried arrays should be stored at –20°C, sealed individually in plastic bags.

3.2. RNA Isolation

3.2.1. RNA Isolation from Tissues

Conical 50-mL tubes can break under forces >10,000*g*. We recommend using sterile 15- and 50-mL round-bottomed, polypropylene centrifuge tubes at all times.

1. Harvest the tissue; use immediately or flash freeze in liquid nitrogen and store at –70°C. *Important:* When working with frozen tissue, be sure to keep the

Table 2
Reagents for RNA Isolation from Tissues

	Weight of tissue			
	10–100 mg	100–300 mg	300–600 mg	0.6–1.0 g
Recommended tube size (mL)	2 × 2	15[a]	50[a]	50[a]
Denaturing solution (mL)	1.0	3.0	6.0	10.0
Saturated phenol (mL)	2.0	6.0	12.0	20.0
Chloroform (mL)	0.6	1.8	3.6	6.0
Saturated phenol (2nd round) (mL)	1.6	4.8	9.6	16.0
Chloroform (2nd round) (mL)	0.6	1.8	3.6	6.0
Isopropanol (mL)	2.0	6.0	12.0	20.0
80% EtOH wash (mL)	1.0	3.0	6.0	10.0

[a]Conical tubes can break under forces greater than 10,000g. Ensure that round-bottomed tubes are used.

tissue frozen until you add the denaturing solution. Even partial thawing can result in RNA degradation. Perform all necessary manipulations on dry ice or liquid nitrogen.
2. Cut or crush the tissue into small pieces (<1 cm^3). When working with frozen tissue, prechill a mortar and pestle with liquid nitrogen, fill the mortar with liquid nitrogen, and break frozen tissue into smaller pieces.
3. Weigh out the tissue in a prechilled, sterile tube. See **Table 2** for the appropriate tube size.
4. Add the appropriate volume (*see* **Table 2**) of denaturing solution. Always add at least 1 mL/100 mg of tissue.
5. Grind the sample at 0–4°C using a tissue homogenizer (e.g., Polytron or equivalent) at the maximum setting for 1 to 2 min or until completely homogenized.
6. Incubate on ice for 5–10 min.
7. Vortex the sample thoroughly. Centrifuge the homogenate at 15,000g for 5 min at 4°C to remove cellular debris.
8. Transfer the entire supernatant to new centrifuge tube(s). Avoid pipeting the insoluble upper layer, if present.
9. Add the appropriate volume (*see* **Table 2**) of saturated phenol.
10. Cap the tubes securely and vortex for 1 min. Incubate on ice for 5 min.
11. Add the appropriate volume (*see* **Table 2**) of chloroform.
12. Shake the sample and vortex vigorously for 1 to 2 min. Incubate on ice for 5 min.
13. Centrifuge the homogenate at 15,000g for 10 min at 4°C.
14. Transfer the upper aqueous phase containing the RNA to a new tube. Take care not to pipet any material from the white interface or lower organic phase.
15. Perform a second round of phenol:chloroform extraction, using the amounts shown in **Table 2** for "2nd round" (*see* **Note 1**). Repeat **steps 9–14**.

Table 3
Reagents for RNA Isolation from Cultured Cells

	Cell number			
	10^6–10^7	1–3 × 10^7	3–6 × 10^7	6–10 × 10^7
Tube size (mL)	2 × 2	15[a]	50[a]	50[a]
Denaturing solution (mL)	1.0	3.0	6.0	10.0
Saturated phenol (mL)	2.0	6.0	12.0	20.0
Chloroform (mL)	0.6	1.8	3.6	6.0
Saturated phenol (2nd round) (mL)	1.6	4.8	9.6	16.0
Chloroform (2nd round) (mL)	0.6	1.8	3.6	6.0
Isopropanol (mL)	2.0	6.0	12.0	20.0
80% EtOH wash (mL)	1.0	3.0	6.0	10.0

[a]Conical tubes can break under forces greater than 10,000g. Ensure that round-bottomed tubes are used.

16. Transfer the upper phase to a new tube. Avoid touching the interface.
17. Add the appropriate volume (*see* **Table 2**) of isopropanol. Add slowly, mixing occasionally as you add it.
18. Mix the solution well and incubate on ice for 10 min.
19. Centrifuge the samples at 15,000g for 15 min at 4°C.
20. Quickly remove the supernatant without disturbing the RNA pellet.
21. Add the appropriate volume (*see* **Table 2**) of 80% ethanol.
22. Centrifuge at 15,000g for 5 min at 4°C. Quickly and carefully discard the supernatant.
23. Air-dry the pellet.
24. Resuspend the pellet in enough RNase-free H_2O to ensure an RNA concentration of 1 to 2 µg/µL. Refer to **Table 4** for approximate yields.
25. Allow the pellet to soak, then resuspend thoroughly by tapping the tube and pipeting.
26. Set aside a 2-µL aliquot to compare with your RNA sample following DNase treatment. Store the RNA samples at –70°C until ready to proceed with DNase treatment.

3.2.2. RNA Isolation from Cultured Cells

1. Transfer the cultured cells to a sterile tube. See **Table 3** for the appropriate tube size.
2. Centrifuge at 500g for 5 min at 4°C. Discard the supernatant.
3. Use the cells immediately, or flash freeze in liquid nitrogen and store at –70°C. When working with frozen cells, be sure to keep the cells frozen until you add

Table 4
Representative Total RNA Yields

Tissue/cell source	Amount of starting material	Yield of total RNA (µg)	Yield after DNase (70% recovery) (µg)
Rat liver	100 mg	600	420
Rat skeletal muscle	100 mg	90	60
Mouse brain	100 mg	125	90
Mouse spleen	100 mg	245	170
Mouse testes	100 mg	240	170
Mouse thymus	100 mg	85	60
Human cerebellum	100 mg	85	60
Human prostate tumor	100 mg	100	70
MCF-7 cell line	1×10^7 cells	70	50
Mouse fibroblasts	1×10^7 cells	800	560
U251 cell line	1×10^7 cells	95	65

the denaturing solution. Even partial thawing can result in RNA degradation. Perform all necessary manipulations on dry ice or liquid nitrogen.
4. Add the appropriate volume (**Table 3**) of denaturing solution.
5. Pipet up and down vigorously and vortex well until the cell pellet is completely resuspended.
6. Incubate on ice for 5–10 min.
7. Vortex the sample thoroughly. Centrifuge the homogenate at 15,000*g* for 5 min at 4°C to remove cellular debris.
8. Transfer the entire supernatant to new centrifuge tube(s). Avoid pipeting the insoluble upper layer, if present.
9. Add the appropriate volume (*see* **Table 3**) of saturated phenol.
10. Cap the tubes securely and vortex for 1 min. Incubate on ice for 5 min.
11. Add the appropriate volume (*see* **Table 3**) of chloroform.
12. Shake the sample and vortex vigorously for 1 to 2 min. Incubate on ice for 5 min.
13. Centrifuge the homogenate at 15,000*g* for 10 min at 4°C.
14. Transfer the upper aqueous phase containing the RNA to a new tube. Take care not to pipet any material from the white interface or lower organic phase.
15. Perform a second round of phenol:chloroform extraction, using the amounts shown in **Table 3** for "2nd round" (*see* **Note 1**). Repeat **steps 9–14**.
16. Transfer the upper phase to a new tube. Avoid touching the interface.
17. Slowly add the appropriate volume (*see* **Table 3**) of isopropanol, mixing occasionally as you add it.
18. Mix the solution well and incubate on ice for 10 min.

19. Centrifuge the samples at 15,000g for 15 min at 4°C.
20. Quickly remove the supernatant without disturbing the RNA pellet.
21. Add the appropriate volume (*see* **Table 3**) of 80% ethanol.
22. Centrifuge at 15,000g for 5 min at 4°C. Quickly and carefully discard the supernatant.
23. Air-dry the pellet.
24. Resuspend the pellet in enough RNase-free H_2O to ensure an RNA concentration of 1 to 2 µg/µL. Refer to **Table 4** for approximate yields.
25. Allow the pellet to soak, and then resuspend thoroughly by tapping the tube and pipeting.
26. Set aside a 2-µL aliquot to compare with your RNA sample following DNase treatment. Store the RNA samples at −70°C until ready to proceed with DNase treatment.

3.2.3. DNase Treatment

The following protocol describes DNase I treatment of 0.5 mg of total RNA prior to purification of poly A^+ RNA. If you are starting with more or less than 0.5 mg, adjust all volumes proportionally.

1. Combine the following reagents in a 1.5-mL microcentrifuge tube for each sample (you may scale up or down accordingly): 500 µL of total RNA (1 mg/mL), 100 µL of 10X DNase I buffer, 50 µL of DNase I (1 U/µL), and 350 µL of deionized H_2O, for a total volume of 1.0 mL. Mix well by pipeting.
2. Incubate the reactions at 37°C for 30 min in an air incubator.
3. Add 100 µL of 10X termination mix. Mix well by pipeting.
4. Split each reaction into two 1.5-mL microcentrifuge tubes (550 µL per tube).
5. Add 500 µL of saturated phenol and 300 µL of chloroform to each tube and vortex thoroughly.
6. Centrifuge at 16,000g for 10 min at 4°C to separate the phases.
7. Carefully transfer the top aqueous layer to a fresh 1.5-mL microcentrifuge tube. Avoid pipeting any material from the interface or lower phase.
8. Add 550 µL of chloroform to the aqueous layer and vortex thoroughly.
9. Centrifuge at 16,000g for 10 min at 4°C to separate the phases.
10. Carefully remove the top aqueous layer and place in a 2-mL microcentrifuge tube.
11. Add 1/10 vol (50 µL) of 2 *M* NaOAc and 2.5 vol (1.5 mL) of 95% ethanol. If treating <20 µg of total RNA, add 20 µg of glycogen.
12. Vortex the mixture thoroughly; incubate on ice for 10 min.
13. Spin in a microcentrifuge at 16,000g for 15 min at 4°C.
14. Carefully remove the supernatant and any traces of ethanol.
15. Gently overlay the pellet with 500 µL of 80% ethanol.
16. Centrifuge at 16,000g for 5 min at 4°C.
17. Carefully remove the supernatant.

18. Air-dry the pellet for approx 10 min or until the pellet is dry.
19. Dissolve the precipitate in 250 µL of RNase-free H_2O, and assess the yield and purity of the RNA as described in **Subheading 3.3.** Alternatively, store the RNA at –70°C.

3.3. Assessment of RNA Yield and Quality (see Table 4)

3.3.1. Calculation of A_{260}/A_{280} Ratio

Pure RNA exhibits a ratio of 1.9–2.1.

3.3.2. Gel Electrophoresis

Electrophorese 1 to 2 µg of total RNA on a 1% denaturing agarose gel. Examine the gel when the dye has migrated 3 to 4 cm from the wells. Total RNA from mammalian sources should appear as two bright bands (28S and 18S RNA) at approx 4.5 and 1.9 kb (*see* **Note 2**). The ratio of intensities of the 28S and 18S rRNA bands should be 1.5–2.5:1. Lower ratios are indicative of degradation. You may also see additional bands or a smear lower than the 18S rRNA band, including very small bands corresponding to 5S rRNA and tRNA.

3.3.3. Testing for DNA Contamination by PCR

A simple test for genomic DNA contamination is to use the total RNA directly as a template in a PCR reaction with primers for any well-characterized gene (e.g., actin or G3PDH). Select primers that will amplify a genomic DNA fragment <1 kb. Be careful that the primers are not separated by a long intron. If this reaction produces bands that are visible on an agarose/ethidium bromide (EtBr) gel, the RNA almost certainly contains genomic DNA. As a positive control, use different concentrations of genomic DNA as a template for PCR. This control will allow you to determine the approximate percentage of DNA impurities in the RNA sample. For a successful nylon array experiment, the RNA should contain <0.001% genomic DNA or produce no visible PCR product after 35 cycles.

3.4. Poly A^+ Enrichment and Preparation of Probes (see Note 3)

3.4.1. Preparation of Streptavidin Magnetic Beads

1. Resuspend magnetic beads by inverting and gently tapping the tube.
2. Aliquot 15 µL of beads per probe synthesis reaction into one 0.5-mL tube.
3. Separate the beads on a magnetic particle separator.
4. Pipet off and discard the supernatant.
5. Wash the beads with 150 µL of 1X binding buffer; pipet up and down.
6. Separate the beads on a magnetic particle separator.
7. Pipet off and discard the supernatant.

8. Repeat **steps 5–7** three times.
9. Resuspend the beads in 15 µL of 1X binding buffer per reaction.

3.4.2. Enrichment of Poly A⁺ RNA

Perform the following steps for each total RNA sample. It is extremely important that you do not pause between any of these steps.

1. Preheat a PCR thermal cycler to 70°C.
2. Aliquot 10–50 µg of total RNA into a 0.5-mL tube. For synthesizing probes with the highest sensitivity, we recommend using as much RNA as possible, up to the 50-µg limit.
3. Add deionized H_2O to 45 µL.
4. Add 1 µL of biotinylated oligo(dT), and thoroughly mix by pipeting.
5. Incubate at 70°C for 2 min in the preheated thermal cycler.
6. Remove from heat and cool at room temperature for 10 min.
7. Add 45 µL of 2X binding buffer, and mix well by pipeting.
8. Resuspend the washed beads by pipeting up and down, and add 15 µL to each RNA sample.
9. Mix on a vortexer or shaker at 1500 rpm for 25–30 min at room temperature. Ensure that the beads remain suspended. Do not exceed 30 min.
10. Separate the beads using the magnetic separator. Carefully pipet off and discard the supernatant.
11. Gently resuspend the beads in 50 µL of 1X wash buffer.
12. Being careful not to lose particles, separate the beads and then pipet off and discard the supernatant.
13. Repeat **steps 11** and **12** one time.
14. Gently resuspend the beads in 50 µL of 1X reaction buffer.
15. Separate the beads, and then pipet off and discard the supernatant.
16. Resuspend the beads in 3 µL of deionized H_2O.

3.4.3. cDNA Probe Synthesis

The generation of cDNA probes from total or poly A⁺ RNA is accomplished through reverse transcription. The reverse transcription reaction can be primed with a random primer mix, or with a gene-specific mix of antisense primers that generates cDNA for only those genes represented on your array (if the array contains less than 3000–4000 genes). We have found that preparing a gene-specific primer mix for each different array results in an approx 10-fold increase in sensitivity, with a concomitant reduction in nonspecific background. To prepare a 10X gene-specific primer mix for your array, prepare a mixture of 25-bp antisense primers representing each gene of the array, with a final, combined DNA concentration for all primers of 30–50 μM.

1. Prepare a master mix for all labeling reactions plus one extra reaction (to ensure that you have sufficient volume). Combine the following (per reaction) in a 0.5-mL microcentrifuge tube at room temperature (see **Note 4**): 4 µL of 5X reaction buffer (see **Note 5**), 2 µL of 10X dNTP mix (for dATP label), 5 µL of [α-^{32}P]dATP (3000 Ci/mmol, 10 µCi/µL) or [α-^{33}P]dATP (>2500 Ci/mmol, 10 µCi/µL), and 0.5 µL of DTT (100 mM), for a total volume of 11.5 µL.
2. Preheat a PCR thermal cycler to 65°C.
3. Add 4 µL of 10X gene-specific primer mix or 4 µL of random primer mix to the resuspended beads. Mix well by pipeting.
4. Incubate the beads and primer mix in the preheated thermal cycler at 65°C for 2 min.
5. Reduce the temperature of the thermal cycler to 50°C (or 48°C if using an unregulated heating block or water bath); incubate the tubes for 2 min. During this incubation, add 2 µL of PowerScript Reverse Transcriptase (or MMLV RT; see **Note 5**) per reaction to the master mix by pipeting, and keep the master mix at room temperature.
6. After completion of the 2-min incubation at 50°C, add 13.5 µL of master mix to each reaction tube. Mix the contents of the tubes thoroughly by pipeting, and immediately return them to the thermal cycler.
7. Incubate the tubes at 50°C (or 48°C) for 25 min.
8. Add 2 µL of 10X termination mix, and mix well.
9. Separate the beads and pipet the supernatant (~approx 20 µL) into 180 µL of Buffer NT2.
10. Place a NucleoSpin extraction spin column into a 2-mL collection tube, and pipet the sample into the column. Centrifuge at 16,000g for 1 min. Discard the collection tube and flowthrough into the appropriate container for radioactive waste.
11. Insert the NucleoSpin column into a fresh 2-mL collection tube. Add 400 µL of buffer NT3 to the column. Centrifuge at 16,000g for 1 min. Discard the collection tube and flowthrough.
12. Repeat **step 11** twice.
13. Transfer the NucleoSpin column to a clean 1.5-mL microcentrifuge tube. Add 100 µL of buffer NE, and allow the column to soak for 2 min.
14. Centrifuge at 14,000 rpm for 1 min to elute the purified probe.
15. Check the radioactivity of the probe by scintillation counting:
    a. Add 2 µL of each purified probe to 5 mL of scintillation fluid in separate scintillation-counter vials.
    b. Count ^{32}P- or ^{33}P-labeled samples on the ^{32}P channel, and calculate the total number of counts in each sample. (Multiply the counts by a dilution factor of 50.) Probes synthesized using this procedure should have a total of $1-10 \times 10^6$ cpm. Store the probes at –20°C.
16. Discard the flowthrough fractions, columns, and elution tubes in the appropriate container for radioactive waste.

3.5. Hybridization to Nylon Arrays

1. Prepare a solution of BD ExpressHyb hybridization solution and sheared salmon testes DNA:
    a. Prewarm 5 mL of hybridization solution at 68°C (*see* **Note 6**).
    b. Heat 0.5 mg of the sheared salmon testes DNA at 95–100°C for 5 min, and then chill quickly on ice.
    c. Mix the heat-denatured sheared salmon testes DNA with the prewarmed hybridization solution. Keep at 68°C until use.
2. Fill a hybridization bottle with deionized H_2O. Wet the nylon array by placing it in a dish of deionized H_2O, and then place the membrane in the bottle. Pour off all the water from the bottle; the membrane should adhere to the inside walls of the container without creating air pockets. Add 5 mL of the solution prepared in **step 1**. Ensure that the solution is evenly distributed over the membrane. Perform this step quickly to prevent the array membrane from drying.
3. Prehybridize for 30 min with continuous agitation at 68°C. Do not remove the nylon array from the container during the prehybridization, hybridization, or washing steps. If performing the hybridization in roller bottles, rotate at 5–7 rpm during the prehybridization and hybridization steps.
4. Prepare the probe for hybridization as follows (see **step 5** for optional method):
    a. Add 5 µL of C_0t-1 DNA to the entire pool of labeled probe.
    b. Incubate the probe in a boiling water bath for exactly 2 min.
    c. Incubate the probe on ice for exactly 2 min.
5. **Optional:** We find that boiling is adequate to denature probes; however, if you prefer an alkaline denaturing procedure, you may use the following steps instead:
    a. Mix approx 100 µL of labeled probe (entire sample)~ and approx 11 µL (or 1/10 total volume) of 10X denaturing solution (1 M NaOH, 10 mM EDTA), for a total volume of~ approx 111 µL.
    b. Incubate at 68°C for 20 min.
    c. Add the following to the denatured probe: approx 115 µL (or 1/2 total volume) of 2X neutralizing solution (1 M NaH_2PO_4, pH 7.0), for a total volume of approx 230 µL.
    d. Continue incubating at 68°C for 10 min.
6. Being careful to avoid pouring the concentrated probe directly on the surface of the membrane, add the mixture prepared in **step 4** directly to the array and prehybridization solution. Make sure that the two solutions are mixed.
7. Hybridize overnight with continuous agitation at 68°C. Be sure that all regions of the membrane are in contact with the hybridization solution at all times. If necessary, add an extra 2 to 3 mL of prewarmed BD ExpressHyb hybridization solution.
8. The next day, prewarm wash solution 1 (2X SSC, 1% SDS) and wash solution 2 (0.1X SSC, 0.5% SDS) at 68°C.
9. Carefully remove the hybridization solution and discard in an appropriate radioactive waste container. Replace with 200 mL of prewarmed wash solution

1. Wash the nylon array for 30 min with continuous agitation at 68°C. Repeat this step three more times. If using roller bottles, fill to 80% capacity and rotate at 12–15 rpm during all wash steps.
10. Perform one 30-min wash in 200 mL of prewarmed wash solution 2 with continuous agitation at 68°C.
11. Perform one final 5-min wash in 200 mL of 2X SSC with agitation at room temperature.
12. Using forceps, remove the nylon array from the container and shake off excess wash solution. Do not blot dry or allow the membrane to dry. If the membrane dries even partially, subsequent removal of the probe (stripping) from the nylon array will be difficult.
13. Immediately wrap the damp membrane in plastic wrap.
14. Mount the plastic-wrapped nylon array on Whatman paper (3 MM Chr). Expose the nylon array to X-ray film at –70°C with an intensifying screen. Try several exposures for varying lengths of time (e.g., 3–6 h, overnight, and 3 d). Alternatively, use a phosphorimager. When exposing the nylon array to a phosphorimaging screen at room temperature, be sure to seal the nylon array membrane in plastic to prevent drying.

3.6. Stripping of Nylon Arrays

To reuse the nylon array after exposure to X-ray film or phosphorimaging, you may remove the cDNA probe by stripping. Perform all steps in a fume hood with appropriate radiation protection.

1. In a 2-L beaker, heat 500 mL of 0.5% SDS solution to boiling.
2. Remove the plastic wrap from the nylon array and immediately place the membrane into the boiling solution. Avoid prolonged exposure of the membrane to air.
3. Continue to boil for 5–10 min.
4. Remove the solution from the heat and allow to cool for 10 min.
5. Rinse the nylon array in wash solution 1 (2X SSC, 1% SDS).
6. Remove the nylon array from the solution and immediately wrap the damp membrane in plastic wrap. Check the efficiency of stripping with a Geiger hand counter and by exposure to X-ray film (*see* **Note 7**). If radioactivity can still be detected, repeat the stripping procedure (**steps 1–5**).
7. Place the nylon array in a hybridization container and proceed with the next hybridization experiment. Alternatively, the nylon array can be sealed and stored in plastic wrap at –20°C until needed. Do not allow the membrane to dry, even partially.

3.7. Interpretation of Results (see Note 8)

3.7.1. Sensitivity of Detection and Background Level

After hybridization and washing, we recommend that you perform a "trial run" exposure (for 3 to 4 h) of the nylon array membranes to X-ray film or a phosphorimaging screen. This will allow you to assess the sensitivity and quality of the hybridization pattern so that you can determine the optimal exposure time for the experiment. For X-ray film, expose the membranes to Kodak BioMax MS film (with the corresponding BioMax MS intensifying screen) at –70°C overnight. In our experience, other X-ray films are two- to fivefold less sensitive than BioMax MS film. If available, a phosphorimager affords approximately the same sensitivity as BioMax MS film and allows you to quantify hybridization signals.

3.7.2. Exposure Time

As long as the RNA is of high quality, the signals corresponding to medium- to high-abundance mRNAs (0.05–0.5% of poly A$^+$ RNA) can be easily detected after several hours or an overnight exposure. Usually, an overnight exposure is not sufficient to reveal hybridization signals from rare- to medium-abundance mRNAs, especially when using ^{33}P-labeled probes. The exact number of hybridization signals depends on the complexity of the experimental RNA sample and the set of printed cDNAs and may differ by severalfold. The practical limit for sensitivity is the level of background generated by nonspecific hybridization of the probe to the membrane. Longer exposure times (>7 d) are useful only if the background level is low. Overexposure is not an issue if using a phosphorimager.

Some samples may produce signals that are similar or even higher in intensity than the abundant housekeeping genes. After an overnight exposure with ^{32}P-labeled probes, you should observe signals for the most abundant housekeeping genes, including ubiquitin, phospholipase A$_2$, α-tubulin, β-actin, and G3PDH. These genes are expressed at about 0.1–0.5% abundance in mammalian tissues or cells and can be used as universal positive controls. Note that the ratio of intensities of signals for different housekeeping genes may differ as much as two- to fivefold for different tissues or cells.

Another important parameter is the level of nonspecific hybridization, or background. After overnight exposure, there generally will not be hybridization with blank regions of the membrane or with any negative DNA controls.

3.7.3. Normalization of Hybridization Signals

The best approach for comparing hybridization signals for different samples is to equalize the intensity of the hybridization signals by adjusting exposure times. If one array is uniformly darker than the other, adjust the exposure time of one array until the overall signal is approximately the same on both arrays. The most common reason for different overall hybridization intensities is the quality of RNA samples used to prepare the hybridization probes. In our experience, it is most effective and convenient to normalize arrays based on the overall signal from all genes on the array.

As an alternative to normalization based on the overall level of signal, some researchers prefer to identify one or more housekeeping genes that generate equally intense hybridization signals for the samples being compared. This housekeeping gene (or genes) can then serve as a standard for normalization. In cells or tissues that are closely related—i.e., where only a few genes change their expression levels—the expression of housekeeping genes generally remains constant. However, the expression levels of individual housekeeping genes may be variable depending on your experimental system, especially if different tissues are being compared.

4. Notes

1. For very RNase-rich samples (e.g., pancreas, liver, spleen), we recommend that you perform a third or fourth round of phenol:chloroform extraction.
2. If, on a denaturing formaldehyde/agarose/EtBr gel, the total RNA appears as a smear that is no larger than 2 kb, the RNA may be degraded. If this is the case, we suggest you prepare fresh RNA after checking the purification reagents for RNase or other impurities. If problems persist, you may need to find another source of tissue/cells.
3. Be sure to work through the enrichment/probe synthesis steps quickly, without pausing. Additionally, to help reduce any chance of RNA degradation, you may add 100 U of Ambion's ANTI-RNase (cat. no. 2692) after adding magnetic beads to the sample.
4. As discussed in the **Subheading 1.**, both ^{32}P- and ^{33}P-labeling methods are compatible with nylon membrane arrays. Compared with ^{32}P, the spatial resolution and quality of images are improved with ^{33}P. These characteristics tend to facilitate image analysis and signal quantification. However, also note that ^{33}P signals are approximately four times less intense, decreasing assay sensitivity.
5. If desired, you may also use the wild-type MMLV RT provided with the BD Atlas Pure Kit; however, you should use the same enzyme to label all probes that will be directly compared. Ensure that you use the correct 5X reaction buffer. For MMLV use 5X MMLV reaction buffer: 250 mM Tris-HCl (pH 8.3), 375 mM KCl, 15 mM MgCl$_2$.
6. Hybridization volume should be increased to 15 mL for large bottles. As a general rule, ensure that there is adequate volume to keep the array thoroughly bathed during the incubation.
7. If you observe high background when reprobing a nylon array, the membrane may not have been stripped completely or may have been allowed to dry. If a membrane is allowed to dry even partially, subsequent removal of the probe will be very challenging. To prevent drying after the final wash, shake off excess solution with forceps (do not blot dry) and immediately wrap the membrane in

plastic wrap or seal it in a polyethylene bag. When reprobing, unwrap the array, immediately place it in stripping solution, and follow the rest of the protocol provided for removing probes.
8. Because of sequence-dependent hybridization characteristics and variations inherent in any hybridization reaction, array data should be considered semiquantitative. We strongly recommend that you corroborate the results of your experiment using RT-PCR.

Reference

1. Duggan, D. J., Bittner, M., Chen, Y., Meltzer, P., and Trent, J. M. (1999) Expression profiling using cDNA microarrays. *Nat. Genet.* **21,** 10–14.

Suggested Readings

Atlas Mouse cDNA Expression Array I (1998) *Clontechniques* **XIII(1),** 2–4.

Chenchik, A., Chen, S., Makhanov, M., and Siebert, P. (1998) Profiling of gene expression in a human glioblastoma cell line using the Atlas Human cDNA Expression Array I. *Clontechniques* **XIII(1),** 16, 17.

DeRisi, J. L., Iyer, V. R., and Brown, P. O. (1997) Exploring the metabolic and genetic control of gene expression on a genomic scale. *Science* **278,** 680–686.

DeRisi, J., Penland, L., Brown, P. O., Bittner, M. L., Meltzer, P. S., Ray, M., Chen, Y., Su, Y. A., and Trent, J. M. (1996) Use of a cDNA microarray to analyse gene expression patterns in human cancer. *Nat. Genet.* **14,** 457–460.

Heller, R. A., Schena, M., Chai, A., Shalon, D., Bedilion, T., Gilmor, J., Wooley, D. E., and Davis, R. W. (1997) Discovery and analysis of inflammatory disease–related genes using cDNA microarrays. *Proc. Natl. Acad. Sci. USA* **94,** 2150–2155.

Hoheisel, J. D. (1997) Oligomer-chip technology. *Trends Biotech.* **15,** 465–469.

Lockhart, D. J., Dong, H., Byrne, M. C., Follettie, M. T., Gallo, M. V., Chee, M. S., Mittmann, M., Wang, C., Kobayashi, M., Horton, H., and Brown, E. L. (1996) Expression monitoring by hybridization to high-density oligonucleotide arrays. *Nat. Biotech.* **14,** 1675–1680.

Nguyen, C., Rocha, D., Granjeaud, S., Baldit, M., Bernard, K., Naquet, P., and Jordan, B. R. (1995) Differential gene expression in the murine thymus assayed by quantitative hybridization of arrayed cDNA clones. *Genomics* **29,** 207–216.

Piétu, G., Alibert, O., Guichard, V., Lamy, B., Bois, F., Leroy, E., Mariage-Samson, R., Houlgatte, R., Soularue, P., and Auffray, C. (1996) Novel gene transcripts preferentially expressed in human muscles revealed by quantitative hybridization of a high density cDNA array. *Genome Res.* **6,** 492–503.

Schena, M. (1996) Genome analysis with gene expression microarrays. *BioEssays* **18(5),** 427–431.

Schena, M., Shalon, D., Heller, R., Chai, A., Brown, P. O., and Davis, R. W. (1996) Parallel human genome analysis: microarray-based expression monitoring of 1000 genes. *Proc. Natl. Acad. Sci. USA* **93,** 10,614–10,619.

Spanakis, E. (1993) Problems related to the interpretation of autoradiographic data on gene expression using common constitutive transcripts as controls. *Nucleic Acids Res.* **21(16),** 3809–3819.

Wodicka, L., Dong, H., Mittmann, M., Ming-Hsiu, H., and Lockhart, A. (1997) Genome-wide expression monitoring in Saccharomyces cerevisiae. *Nat. Biotech.* **15,** 1359–1367.

Zhang, W., Chenchik, A., Chen, S., Siebert, P., and Rhee, C. H. (1997) Molecular profiling of human gliomas by cDNA expression array. *J. Genet. Med.* **1,** 57–59.

Zhao, N., Hashida, H., Takahashi, N., Misumi, Y., and Sakaki, Y. (1995) High-density cDNA filter analysis: A novel approach for large-scale quantitative analysis of gene expression. *Gene* **166,** 207–213.

3

Plastic Microarrays

A Novel Array Support Combining the Benefits of Macro- and Microarrays

Alexander Munishkin, Konrad Faulstich, Vissarion Aivazachvili, Claire Granger, and Alex Chenchik

1. Introduction

Until recently, gene arrays could only be printed on two types of supports: nylon membranes or glass slides. Nylon membrane–based arrays allow researchers to analyze hundreds of genes in a single experiment using standard laboratory equipment. However, the density of genes that can be included on a membrane array is limited by the printing resolution. Glass arrays allow for a high printing density but can be expensive and require the use of specialized equipment. In response, BD Biosciences Clontech has developed a new type of array support, the BD Atlas™ Plastic Film, which combines the simplicity of nylon arrays with the high gene density of glass arrays.

1.1. Key Properties of Atlas Plastic Support

The plastic format has many advantages that make analysis easy and accurate. Plastic is nonporous, like glass arrays, which allows printing with great precision, resulting in a higher gene density than is possible with a nylon membrane (**Fig. 1**). The nonporous nature of the plastic surface also greatly decreases nonspecific binding, producing a clean background with minimal washing. Radioactive detection is used for plastic arrays, like nylon arrays, enabling the most sensitive detection of gene expression *(1)* across the widest dynamic range. Also like nylon arrays, plastic arrays require no special equip-

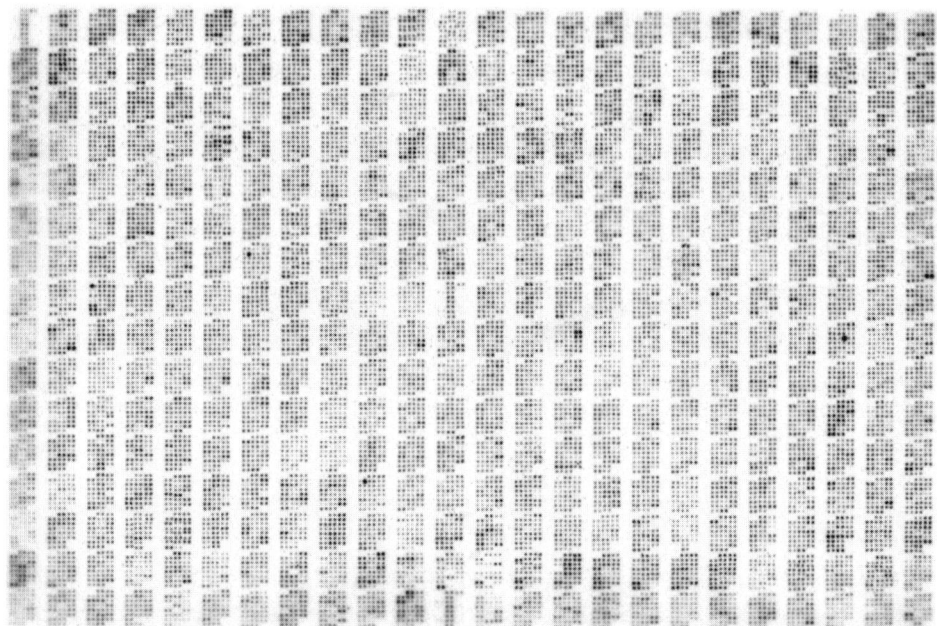

Fig. 1. Hybridization of ^{33}P-labeled antisense oligos to the BD Atlas™ Plastic Human 8K Microarray.

ment for imaging (just a standard phosphorimager). However, Atlas Plastic Film is rigid and does not warp when exposed to high washing temperatures, so it can be stripped and reprobed several times and still retain its original configuration. Since the film does not distort, image analysis is also much easier than analysis of data from nylon membrane–based arrays. Additionally, the spots on the film's surface yield a uniform distribution of signal intensity, which is critical for accurate automated data processing.

The BD Atlas Plastic Film is a clear, rigid sheet that requires no further treatment before printing and is stable at room temperature for at least a year. The BD Atlas™ Plastic Printing Buffer, included in the BD Atlas Plastic Printing Kit™, is specifically designed to produce small, uniform spots of a high printing density when used with the Plastic Film. The Plastic Printing Buffer exhibits minimal evaporation at room temperature, so there is little change in nucleic acid concentration over the course of the printing run. Following UV crosslinking, the components of the Plastic Printing Buffer simply wash away with water, ensuring that nothing interferes with subsequent hybridization experiments.

1.2. Binding Capacity and Binding Efficiency

The binding capacity of a given array support is defined as the maximum amount of nucleic acid that can be bound. The binding capacity of BD Atlas Plastic Film is approx 20 fmol/mm^2. Binding efficiency can be determined either as (1) the amount of nucleic acid that remains bound to the surface (after washing) compared with the amount initially applied, or as (2) the amount of nucleic acid that remains bound to the surface (after washing) relative to the binding capacity. Binding efficiency, relative to the binding capacity of the surface, is about 85–90%, when using the protocol given herein. This capacity and efficiency result in an acceptable compromise between high signal intensities on the array and cost-efficient production.

1.3. Hybridization Time and Efficiency

Since BD Atlas Plastic Film is nonporous, the time required for hybridization is relatively short. Hybridization kinetics are slower for nylon-membrane arrays since the probe must diffuse through the membrane pores. In fact, our experimental results have shown that efficient hybridization to plastic arrays can be achieved in as little as 3 h. Because of the low background of the plastic arrays (low nonspecific adsorption of labeled probe to the surface), the time required for washing procedures is minimized as well. Thus, an entire experimental procedure can be performed in a single day, from probe labeling through hybridization, washing, and, finally, detection of results.

1.4. Array Quality and Calibration

At BD Biosciences Clontech, we have devised methods to test the quality and reproducibility of data derived from plastic array hybridization experiments. You may wish to incorporate one or both test strategies into the design of your plastic array printing and analysis protocol.

In our experience, plastic arrays printed with long (60–80 base) oligonucleotides, "long oligos," yield superior sensitivity (**Figs. 2** and **3**). To ensure quality, every oligo printed on our premade BD Atlas Plastic Microarrays is thoroughly tested to confirm its identity and its ability to produce a strong, specific hybridization signal. This analysis consists of antisense hybridization experiments. We use a mixture of antisense oligos corresponding to each oligo from a particular section of the microarray. These antisense oligos are radiolabeled and hybridized to the entire microarray (**Fig. 1**). Oligos that display weak hybridization signals or visible levels of hybridization to other fragments on the microarray are redesigned. This test ensures that each oligo is capable of producing a strong, specific hybridization signal. Our studies

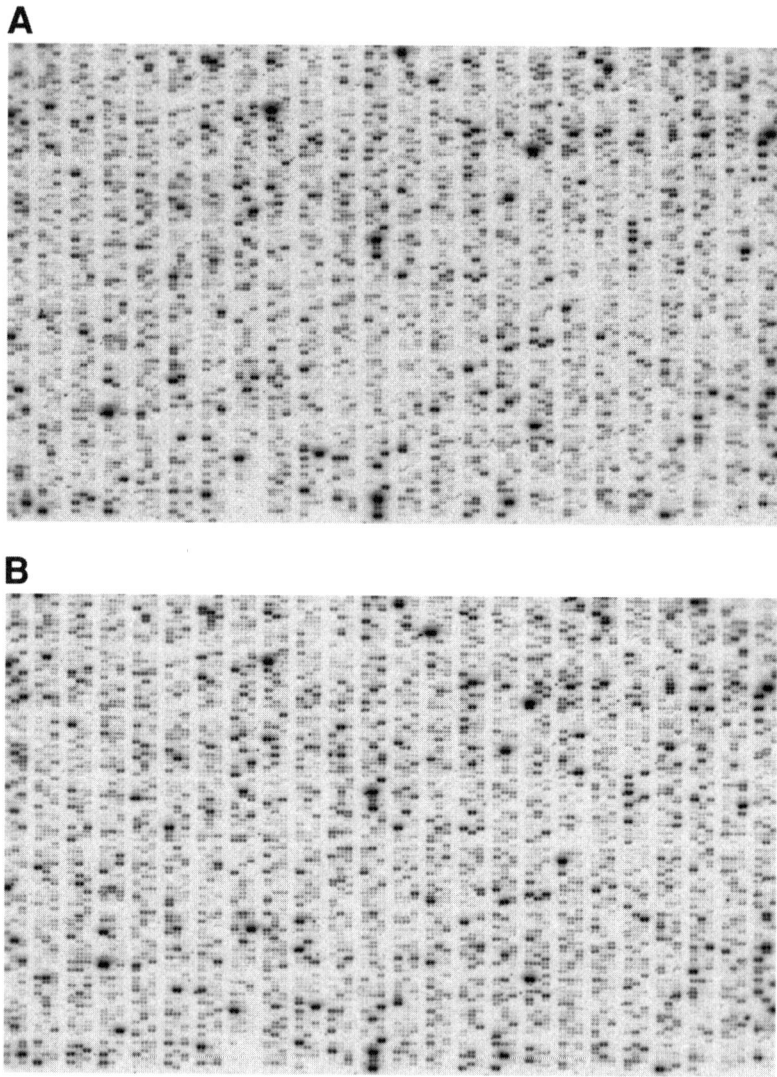

Fig. 2. Expression profiling of normal and diabetic human skeletal muscle using BD Atlas Plastic Human 8K Microarrays. Total RNA was isolated from the skeletal muscle of a normal (**A**) or diabetic (**B**) individual. Ten micrograms of total RNA was used to synthesize ^{33}P-labeled cDNA probes using the BD Atlas Pure Total RNA Labeling System (no. K1038-1). Probes were hybridized to separate BD Atlas Plastic Microarrays. Microarrays were analyzed by phosphorimaging.

indicate that without this test, about 25% of all arrayed oligos would not produce a usable hybridization signal. This test also confirms that each oligo was synthesized correctly and is arrayed in the correct location.

Because of the small variations that are inherent in any array manufacturing process, comparisons of results generated using arrays that were printed at different times have only limited validity. To solve this problem, each lot of our premade BD Atlas Plastic Microarrays is individually calibrated, making it possible to compare microarrays from different lots. This procedure can easily be performed in other array printing facilities. When each new lot of microarrays is printed, several microarrays (three from the beginning, middle, and end of the printing run) are hybridized with a radiolabeled antisense oligo mixture corresponding to all of the genes on the array. The hybridization intensity for each oligo on each microarray is determined, and the values are averaged for each gene across the tested microarrays to produce a set of calibration standards for that lot. This set of calibration standards contains a calibration factor for each gene on the array. To compare hybridization results for microarrays from different lots, one simply multiplies the hybridization intensity for each microarray gene by its factor in the corresponding lot-specific set of calibration standards. Note, however, that because microarray stripping may slightly alter the oligo content on the plastic, calibration standards should not be used to adjust intensity values for reprobed arrays.

1.5. General Considerations

The following protocol is a general description of the printing process. Because of the variety of arraying devices and pin tools currently available, these steps must be optimized for different systems. This protocol is intended for printing long oligonucleotide (60–80 bases) arrays. For best results, the protocol should be optimized to reflect the length and type of nucleic acids being printed. It may be necessary to modify nucleic acid concentration and/or UV crosslinking energy. This protocol has not yet been optimized for printing cDNA arrays.

2. Materials

Unless otherwise noted, all catalog numbers provided are for BD Biosciences Clontech products.

2.1. Microarray Printing Reagents

2.1.1. Reagents in BD Atlas Plastic Printing Kit

The kit is available exclusively from BD Biosciences Clontech (cat. no. K1846-1).

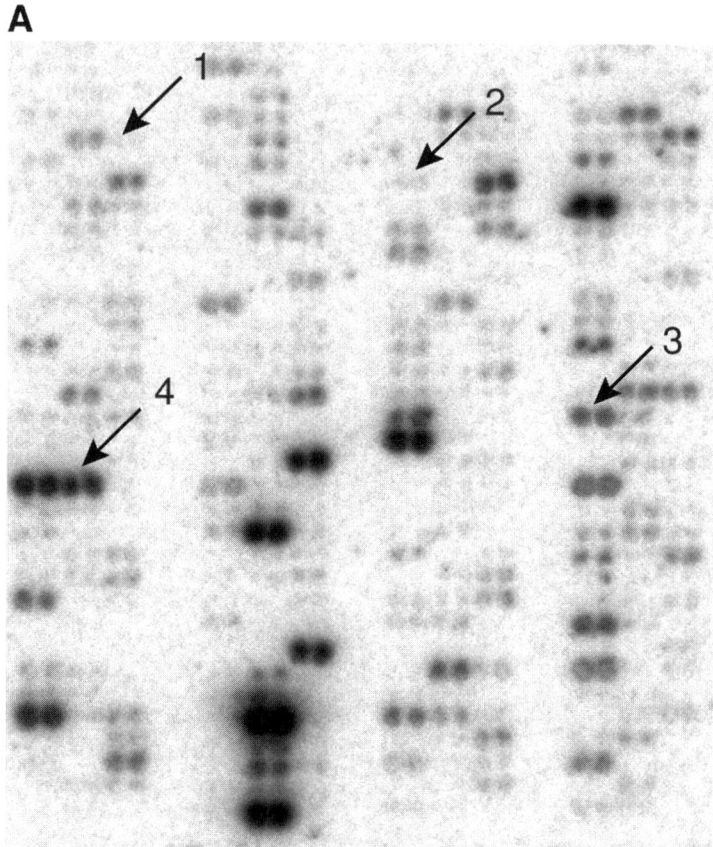

Fig. 3. Differential gene expression detected with BD Atlas Plastic Human 8K Microarrays. The images shown are close-up details of the microarrays shown in **Fig. 2**. Arrows 1, 2, and 5 indicate genes that are upregulated in normal tissue (**A**) compared with diabetic tissue (**B**). Arrows 3 and 4 indicate genes that are downregulated in diabetic tissue compared with normal.

1. BD Atlas Plastic Films. Dimensions of each film: 8.25 cm × 12.22 cm × 178 µm.
2. 8X BD Atlas Plastic Printing Buffer.

2.1.2. Additional Reagents/Special Equipment

1. Sterile, nuclease-free Milli-Q H_2O.
2. Source plate (we recommend Corning 96- or 384-well V-bottomed microplates).
3. Adhesive foil sheets for sealing source plate (we recommend Biomek Seal & Sample Aluminum Foil Lids, Beckman Coulter, cat. no. 538619).
4. Microarray printing machine.

Plastic Microarrays

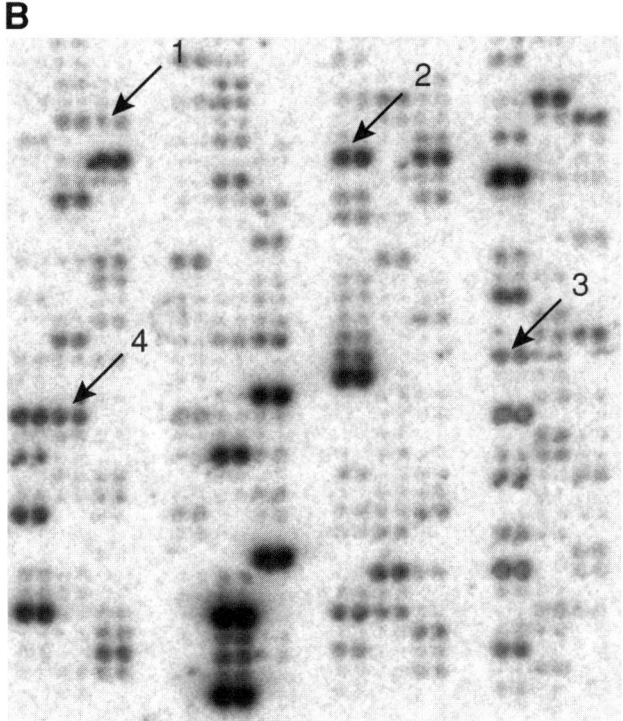

Fig. 3.

5. Microarray pins and pin tool (we recommend using solid pins).
6. Adhesive tape.
7. **Optional:** Microseal A Film (cat. no. MSA-5001; MJ Research).
8. UV crosslinker (we recommend Stratagene's Stratalinker 2400, cat. no. 400075).
9. Lint-free tissues (KimWipes Lab Wipes, cat. no. Z188965, Sigma, St. Louis, MO).

2.2. Reagents for RNA Isolation and Probe Synthesis

2.2.1. Reagents Provided with BD Atlas™ Pure Total RNA Labeling System

The labeling system is available exclusively from BD Biosciences Clontech (cat. no. K1038-1).

1. Denaturing solution.
2. Saturation buffer for phenol.
3. RNase-free H_2O.
4. 2 M NaOAc (pH 4.5).

5. 10X Termination mix.
6. Streptavidin magnetic beads.
7. 1X Binding buffer.
8. 2X Binding buffer.
9. 1X Reaction buffer.
10. 1X Wash buffer.
11. DNase I (1 U/µL).
12. 10X DNase I buffer.
13. Biotinylated oligo(dT).

2.2.2. Additional Reagents/Special Equipment

1. Saturated phenol (store at 4°C). For 160 mL: 100 g of phenol (Sigma cat. no. P1037 or Boehringer Mannheim cat. no. 100728). In a fume hood, heat a jar of phenol in a 70°C water bath for 30 min or until the phenol is completely melted. Add 95 mL of phenol directly to the saturation buffer and mix well. Hydroxyquinoline may be added if desired. Aliquot and freeze at –20°C for long-term storage. This preparation of saturated phenol will have only one phase.
2. Chloroform (cat. no. C2432 or cat. no. C0549; Sigma).
3. Isopropanol (cat. no. I9516; Sigma).
4. Liquid nitrogen or dry ice.
5. Tissue homogenizer (e.g., Polytron or equivalent). For <200 mg of tissue, use a 6-mm probe; for >200 mg of tissue, use a 10-mm probe.
6. [α-^{33}P]dATP (10 µCi/µL; >2500 Ci/mmol) (cat. no. BF1001; Amersham). Do not use Amersham's Redivue or any other dye-containing isotope. ^{32}P-labeling is not compatible with plastic arrays printed at high density.
7. Ethanol (reagent grade).
8. Deionized H_2O (Milli-Q-filtered or equivalent; do not use diethylpyrocarbonate-treated H_2O).
9. Magnetic particle separator (cat. no. Z5331; Promega, Madison, WI). It is important that you use a separator designed for 0.5-mL tubes.
10. Polypropylene centrifuge tubes: 1.5-mL (cat. no.72-690-051; Sarstedt), 2-mL (cat. no. 16-8105-75; PGC), 15-mL (tubes cat. no. 05-562-10D; caps cat. no. 05-562-11E; Fisher), and 50-mL (tubes with caps cat. no. 05-529-1D; Fisher). Fifteen-mL and 50-mL tubes should be sterilized with 1% sodium dodecyl sulfate (SDS) and ethanol before use.
11. 10X dNTP mix (for dATP label) (40 µM dATP; 5 mM each of dCTP, dGTP, dTTP).
12. 10X Random primer mix (N-15).
13. BD PowerScript™ Reverse Transcriptase and 5X BD PowerScript Reaction Buffer (available exclusively from BD Biosciences Clontech; cat. no. 8460-1). If desired, you may also use the wild-type Moloney murine leukemia virus reverse transcriptase (MMLV RT) provided with the BD Atlas Pure Kit; however, you should use the same enzyme to label all probes that will be directly compared.

Ensure that you use the correct 5X reaction buffer. For MMLV use 5X MMLV reaction buffer: 250 mM Tris-HCl (pH 8.3), 375 mM KCl, 15 mM MgCl$_2$.
14. Dithiothreitol (DTT) (100 mM).
15. NucleoSpin® Extraction Kit (cat. no. K3051-1): NucleoSpin extraction spin columns, 2-mL collection tubes, buffer NT2, buffer NT3 (add 95% ethanol before use as specified on the label), buffer NE.

2.3. Reagents for Hybridization and Stripping of Plastic Microarrays

1. BD PlasticHyb™ hybridization solution (available exclusively from BD Biosciences Clontech; cat. no. 8017-1).
2. BD Atlas™ Plastic Array Hybridization Box (cat. no. 7930-1): Alternatively, any plastic box that closely matches the dimensions of the plastic film may be used.
3. **Optional:** 10X denaturing solution (1 M NaOH, 10 mM EDTA) (*see* **Subheading 3.5.**).
4. **Optional:** 2X neutralizing solution (1 M NaH$_2$PO$_4$ [pH 7.0]): 27.6 g of NaH$_2$PO$_4$•H$_2$O. Add 190 mL of H$_2$O, adjust the pH to 7.0 with 10 N NaOH if necessary, and add H$_2$O to 200 mL. Store at room temperature (*see* **Subheading 3.5.**).
5. 20X Saline sodium citrate (SSC): 175.3 g of NaCl, 88.2 g of Na$_3$citrate•2H$_2$O. Add 900 mL of H$_2$O, adjust the pH to 7.0 with 1 M HCl if necessary, and add H$_2$O to 1 L. Store at room temperature.
6. 20% SDS: 200 g of SDS. Add H$_2$O to 1 L and heat to 65°C to dissolve. Store at room temperature.
7. High-salt wash solution: 2X SSC, 0.1% SDS.
8. Low salt wash solution 1: 0.1X SSC, 0.1% SDS.
9. Low salt wash solution 2: 0.1X SSC.
10. 0.1 M Na$_2$CO$_3$. (No pH adjustment is required; ensure that you do not inadvertently use NaHCO$_3$.)

3. Methods
3.1. Printing of Plastic Arrays

The BD Atlas Plastic Film can be used to print single-stranded nucleic acid (oligonucleotide or RNA) microarrays. When constructing the plastic microarrays produced at BD Biosciences Clontech, we have found that "long oligos," 60–80 base oligonucleotides, produce the best results. Oligos in this size range combine the high hybridization efficiency of a cDNA fragment with a short oligo's ability to distinguish homologous genes. Long oligos are perfect for higher-density expression profiling experiments in which it becomes necessary to array highly homologous genes.

Each BD Atlas Plastic Film can be printed with as many as 40,000 target oligos. Depending on your experimental needs, you may wish to include negative controls, such as genomic DNA, phage lambda DNA, or yeast DNA, depending on your system. Some researchers also choose to include oligos representing certain housekeeping genes, known to be highly expressed in their experimental samples, to serve as positive controls.

3.1.1. Preparation of Printing Mixture

1. In individual wells of the source plate, dilute each oligo to a total volume of 21 µL using sterile, nuclease-free H_2O. The final concentration of each oligo should be in the range of 10–50 µM.
2. Add 3 µL of 8X BD Atlas Plastic Printing Buffer to each well of the source plate. Mix by carefully pipeting up and down. Alternatively, seal the plate with adhesive aluminum foil, and then vortex carefully (we recommend using a benchtop vortex mixer with a flat pad). Centrifuge the plate briefly in a benchtop centrifuge to collect liquid at the bottom of the wells. This volume is sufficient for printing at least 100 arrays (or up to 300 arrays using 0.2-µm solid pins).

3.1.2. Printing of BD Atlas Plastic Film

If necessary, the BD Atlas Plastic Film can be trimmed or cut to size before printing.

1. Remove the protective sheet from the surface of the BD Atlas Plastic Film (*see* **Note 1**). The film must be handled with care from this point onward, in order to avoid scratching or otherwise damaging the film.
2. Place the BD Atlas Plastic Film in your arraying device, along with the source plate. If necessary, remove the seal from the source plate. Ensure that the dull (nonshiny) side of the BD Atlas Plastic Film is facing up. The printing surface is facing up when the imprinted code can be read. Plastic films must be fixed firmly to a solid support before printing. This can be accomplished by taping each film down with adhesive tape, ensuring that the tape does not overlap the edge of the BD Atlas Plastic Film by more than 5 mm (*see* **Note 2**). We recommend that films be fixed to a tray that is small enough to be transferred subsequently to a UV crosslinker. If necessary, the film can be fixed directly to the floor of the arraying device.
3. Begin the printing process according to the manufacturer's instructions. Humidity adjustments are not critical since the BD Atlas Plastic Printing Buffer exhibits minimal evaporation at room temperature.
4. Once printing is complete, remove the printed film carefully from the arrayer, keeping the array perfectly horizontal. If the film is fixed to the floor of the arraying device, remove the adhesive tape from the film with extreme care, such that the array is not disturbed. The freshly printed spots remain liquid, so it is critical that the newly printed array not be bent, twisted, or tilted to prevent displacement of the individual spots.
5. Place the plastic array in a UV crosslinker, with the printed side facing up. Crosslink using an energy of 150 mJ/cm^2 (1500 × 100 µJ/cm^2; this is roughly equivalent to 25–45 s for most UV crosslinking devices). If the tray of the arrayer

Table 1
Reagents for RNA Isolation from Tissues

	Weight of tissue			
	10–100 mg	100–300 mg	300–600 mg	0.6–1.0 g
Recommended tube size (mL)	2 × 2	15	50	50
Denaturing solution (mL)	1.0	3.0	6.0	10.0
Saturated phenol (mL)	2.0	6.0	12.0	20.0
Chloroform (mL)	0.6	1.8	3.6	6.0
Saturated phenol (2nd round) (mL)	1.6	4.8	9.6	16.0
Chloroform (2nd round) (mL)	0.6	1.8	3.6	6.0
Isopropanol (mL)	2.0	6.0	12.0	20.0
80% EtOH wash (mL)	1.0	3.0	6.0	10.0

is small enough, the entire tray plus the plastic array should be placed in the UV crosslinker.
6. Remove the plastic array from the UV crosslinker. If still fixed to the tray, carefully detach the array. Rinse the film in sterile H_2O for 20 s, using gentle agitation.
7. Remove the array slowly from the H_2O. Let as much liquid as possible drain away before placing the array flat on lint-free tissues to air-dry at room temperature for 10–30 min.
8. Once the array is dry, store at room temperature, away from light, until you are ready to proceed with hybridization.

3.2. RNA Isolation

3.2.1. RNA Isolation from Tissues

Conical 50-mL tubes can break under forces >10,000*g*. We recommend using sterile 15- and 50-mL round-bottomed polypropylene centrifuge tubes at all times.

1. Harvest the tissue; use immediately or flash freeze in liquid nitrogen and store at –70°C. *Important:* When working with frozen tissue, be sure to keep the tissue frozen until you add the denaturing solution. Even partial thawing can result in RNA degradation. Perform all necessary manipulations on dry ice or liquid nitrogen.
2. Cut or crush the tissue into small pieces (<1 cm^3). When working with frozen tissue, prechill a mortar and pestle with liquid nitrogen, fill the mortar with liquid nitrogen, and break frozen tissue into smaller pieces.
3. Weigh out the tissue in a prechilled, sterile tube. See **Table 1** for the appropriate tube size.
4. Add the appropriate volume (*see* **Table 1**) of denaturing solution. Ensure that you add at least 1 mL/100 mg of tissue.

5. Grind the sample at 0–4°C using a tissue homogenizer (e.g., Polytron or equivalent) at the maximum setting for 1 to 2 min or until completely homogenized.
6. Incubate on ice for 5–10 min.
7. Vortex the sample thoroughly. Centrifuge the homogenate at 15,000g for 5 min at 4°C to remove cellular debris.
8. Transfer the entire supernatant to a new centrifuge tube(s). Avoid pipeting the insoluble upper layer, if present.
9. Add the appropriate volume (*see* **Table 1**) of saturated phenol.
10. Cap the tubes securely and vortex for 1 min. Incubate on ice for 5 min.
11. Add the appropriate volume (*see* **Table 1**) of chloroform.
12. Shake the sample and vortex vigorously for 1 to 2 min. Incubate on ice for 5 min.
13. Centrifuge the homogenate at 15,000g for 10 min at 4°C.
14. Transfer the upper aqueous phase containing the RNA to a new tube. Take care not to pipet any material from the white interface or lower organic phase.
15. Perform a second round of phenol:chloroform extraction, using the amounts shown in **Table 1** for "2nd round" (*see* **Note 3**). Repeat **steps 9–14**.
16. Transfer the upper phase to a new tube. Avoid touching the interface.
17. Add the appropriate volume (*see* **Table 1**) of isopropanol. Add slowly, mixing occasionally as you add it.
18. Mix the solution well and incubate on ice for 10 min.
19. Centrifuge the samples at 15,000g for 15 min at 4°C.
20. Quickly remove the supernatant without disturbing the RNA pellet.
21. Add the appropriate volume (*see* **Table 1**) of 80% ethanol.
22. Centrifuge at 15,000g for 5 min at 4°C. Quickly and carefully discard the supernatant.
23. Air-dry the pellet.
24. Resuspend the pellet in enough RNase-free H_2O to ensure an RNA concentration of 1 to 2 µg/µL. Refer to **Table 3** for approximate yields.
25. Allow the pellet to soak, and then resuspend thoroughly by tapping the tube and pipeting.
26. Set aside a 2-µL aliquot to compare with the RNA sample following DNase treatment (*see* **Subheading 3.2.3.**). Store the RNA samples at –70°C until ready to proceed with DNase treatment (*see* **Note 4**).

3.2.2. RNA Isolation from Cultured Cells

Conical 50-mL tubes can break under forces >10,000g. We recommend using sterile 15- and 50-mL round-bottomed polypropylene centrifuge tubes at all times.

1. Transfer the cultured cells to a sterile tube. *See* **Table 2** for the appropriate tube size.
2. Centrifuge at 500g for 5 min at 4°C. Discard the supernatant.

Table 2
Reagents for RNA Isolation from Cultured Cells

	Cell number			
	10^6–10^7	1–3 × 10^7	3–6 × 10^7	6–10 × 10^7
Tube size (mL)	2 × 2	15	50	50
Denaturing solution (mL)	1.0	3.0	6.0	10.0
Saturated phenol (mL)	2.0	6.0	12.0	20.0
Chloroform (mL)	0.6	1.8	3.6	6.0
Saturated phenol (2nd round) (mL)	1.6	4.8	9.6	16.0
Chloroform (2nd round) (mL)	0.6	1.8	3.6	6.0
Isopropanol (mL)	2.0	6.0	12.0	20.0
80% EtOH wash (mL)	1.0	3.0	6.0	10.0

3. Use the cells immediately, or flash freeze in liquid nitrogen and store at –70°C. *Important:* When working with frozen cells, be sure to keep the cells frozen until you add the denaturing solution. Even partial thawing can result in RNA degradation. Perform all necessary manipulations on dry ice or liquid nitrogen.
4. Add the appropriate volume (*see* **Table 2**) of denaturing solution.
5. Pipet up and down vigorously and vortex well until the cell pellet is completely resuspended.
6. Incubate on ice for 5–10 min.
7. Vortex well. Centrifuge the homogenate at 15,000*g* for 5 min at 4°C to remove cellular debris.
8. Transfer the entire supernatant to a new centrifuge tube(s). Avoid pipeting the insoluble upper layer, if present.
9. Add the appropriate volume (*see* **Table 2**) of saturated phenol.
10. Cap the tubes securely and vortex for 1 min. Incubate on ice for 5 min.
11. Add the appropriate volume (*see* **Table 2**) of chloroform.
12. Shake the sample and vortex vigorously for 1 to 2 min. Incubate on ice for 5 min.
13. Centrifuge the homogenate at 15,000*g* for 10 min at 4°C.
14. Transfer the upper aqueous phase containing the RNA to a new tube. Take care not to pipet any material from the white interface or lower organic phase.
15. Perform a second round of phenol:chloroform extraction, using the amounts shown in **Table 2** for "2nd round" (*see* **Note 3**). Repeat **steps 9–14**.
16. Transfer the upper phase to a new tube. Avoid touching the interface.
17. Slowly add the appropriate volume (*see* **Table 2**) of isopropanol, mixing occasionally as you add it.
18. Mix the solution well and incubate on ice for 10 min.
19. Centrifuge the sample at 15,000*g* for 15 min at 4°C.
20. Quickly remove the supernatant without disturbing the RNA pellet.

Table 3
Representative Total RNA Yields

Tissue/cell source	Amount of starting material	Yield of total RNA (µg)	Yield after DNase (70% recovery) (µg)
Rat liver	100 mg	600	420
Rat skeletal muscle	100 mg	90	60
Mouse brain	100 mg	125	90
Mouse spleen	100 mg	245	170
Mouse testes	100 mg	240	170
Mouse thymus	100 mg	85	60
Human cerebellum	100 mg	85	60
Human prostate tumor	100 mg	100	70
MCF-7 cell line	1×10^7 cells	70	50
Mouse fibroblasts	1×10^7 cells	800	560
U251 cell line	1×10^7 cells	95	65

21. Add the appropriate volume (see **Table 2**) of 80% ethanol.
22. Centrifuge at 15,000g for 5 min at 4°C. Quickly and carefully discard the supernatant.
23. Air-dry the pellet.
24. Resuspend the pellet in enough RNase-free H_2O to ensure an RNA concentration of 1 to 2 µg/µL. Refer to **Table 3** for approximate yields.
25. Allow the pellet to soak, and then resuspend thoroughly by tapping the tube and pipeting.
26. Set aside a 2-µL aliquot to compare with the RNA sample following DNase treatment. Store RNA samples at –70°C until ready to proceed with DNase treatment (see **Note 4**).

3.2.3. DNase Treatment

DNase treatment of RNA may improve the quality of results for some samples. We recommend performing this procedure to troubleshoot poor hybridization results or if you suspect your samples are contaminated with DNA. The following protocol describes DNase I treatment of 0.5 mg of total RNA prior to purification of poly A^+ RNA. If you are starting with more or less than 0.5 mg, adjust all volumes proportionally.

1. Combine the following reagents in a 1.5-mL microcentrifuge tube for each sample (you may scale up or down accordingly): 500 µL of total RNA (1 mg/mL), 100 µL of 10X DNase I buffer, 50 µL of DNase I (1 U/µL), and 350 µL of deionized H_2O, for a total volume of 1.0 mL. Mix well by pipeting.
2. Incubate the reactions at 37°C for 30 min in an air incubator.

3. Add 100 µL of 10X termination mix and mix well by pipeting.
4. Split each reaction into two 1.5-mL microcentrifuge tubes (550 µL/tube).
5. Add 500 µL of saturated phenol and 300 µL of chloroform to each tube and vortex thoroughly.
6. Centrifuge at 16,000g for 10 min at 4°C to separate the phases.
7. Carefully transfer the top aqueous layer to a fresh 1.5-mL microcentrifuge tube. Avoid pipeting any material from the interface or lower phase.
8. Add 550 µL of chloroform to the aqueous layer and vortex thoroughly.
9. Centrifuge at 16,000g for 10 min at 4°C to separate the phases.
10. Carefully remove the top aqueous layer and place in a 2.0-mL microcentrifuge tube.
11. Add 1/10 vol (50 µL) of 2 M NaOAc and 2.5 vol (1.5 mL) of 95% ethanol. If treating <20 µg of total RNA, add 20 µg of glycogen.
12. Vortex the mixture thoroughly; incubate on ice for 10 min.
13. Spin in a microcentrifuge at 16,000g for 15 min at 4°C.
14. Carefully remove the supernatant and any traces of ethanol.
15. Gently overlay the pellet with 500 µL of 80% ethanol.
16. Centrifuge at 16,000g for 5 min at 4°C.
17. Carefully remove the supernatant.
18. Air-dry the pellet for approx 10 min or until the pellet is dry.
19. Dissolve the precipitate in 250 µL of RNase-free H$_2$O and assess the yield and purity of the RNA as described in **Subheading 3.3.** Alternatively, store the RNA at –70°C.

3.3. Assessment of RNA Yield and Quality (see Table 3)

3.3.1. Calculation of the A_{260}/A_{280} Ratio

Pure RNA solutions exhibit a ratio of 1.9–2.1.

3.3.2. Gel Electrophoresis

Electrophorese 1 to 2 µg of total RNA on a 1% denaturing agarose gel. Examine the gel when the dye has migrated 3 to 4 cm from the wells. Total RNA from mammalian sources should appear as two bright bands (28S and 18S RNA) at approx 4.5 and 1.9 kb (*see* **Note 5**). The ratio of intensities of the 28S and 18S rRNA bands should be 1.5–2.5:1. Lower ratios indicate degradation. You may also see additional bands or a smear lower than the 18S rRNA band, including very small bands corresponding to 5S rRNA and tRNA.

3.3.3. Testing for DNA Contamination by Polymerase Chain Reaction

A simple test for genomic DNA contamination is to use the total RNA directly as a template in a polymerase chain reaction (PCR) reaction with primers for any well-characterized gene (e.g., actin or G3PDH). Select primers that will amplify a genomic DNA fragment <1 kb. Be careful that the primers are not separated by a long intron. If this reaction produces bands that are visible on an agarose/ethidium bromide (EtBr) gel, the RNA almost certainly contains genomic DNA. As a positive control, use different concentrations of genomic DNA as a template for PCR. This control will allow you to determine the approximate percentage of DNA impurities in the RNA sample. For a successful experiment, the RNA should contain <0.001% genomic DNA or produce no visible PCR product after 35 cycles.

3.4. Poly A⁺ Enrichment and Probe Synthesis (see Note 6)

3.4.1. Preparation of Streptavidin Magnetic Beads

1. Resuspend the magnetic beads by inverting and gently tapping the tube.
2. Aliquot 15 µL of beads per probe synthesis reaction into one 0.5-mL tube.
3. Separate the beads on a magnetic particle separator.
4. Pipet off and discard the supernatant.
5. Wash the beads with 150 µL of 1X binding buffer; pipet up and down.
6. Separate the beads on the magnetic particle separator.
7. Pipet off and discard the supernatant.
8. Repeat **steps 5–7** three times.
9. Resuspend the beads in 15 µL of 1X binding buffer per reaction.

3.4.2. Poly A⁺ RNA Enrichment

It is extremely important that you do not pause between any of these steps.

1. Preheat a PCR thermal cycler to 70°C.
2. Aliquot 10–50 µg of total RNA into a 0.5-mL tube (*see* **Note 7**).
3. Add deionized H₂O to 45 µL.
4. Add 1 µL of biotinylated oligo(dT), and then thoroughly mix by pipeting.
5. Incubate at 70°C for 2 min in the preheated thermal cycler.
6. Remove from heat and cool at room temperature for 10 min.
7. Add 45 µL of 2X binding buffer and mix well by pipeting.
8. Resuspend the washed beads by pipeting up and down, and add 15 µL to each RNA sample.
9. Mix on a vortexer or shaker at 1500 rpm for 25–30 min at room temperature. Ensure that the beads remain suspended. Do not exceed 30 min.
10. Separate the beads using the magnetic separator. Carefully pipet off and discard the supernatant.
11. Gently resuspend the beads in 50 µL of 1X wash buffer.
12. Being careful not to lose particles, separate the beads and then pipet off and discard the supernatant.
13. Repeat **steps 11** and **12** one time.
14. Gently resuspend the beads in 50 µL of 1X reaction buffer.

15. Separate the beads and then pipet off and discard the supernatant.
16. Resuspend the beads in 3 µL of dH$_2$O.

3.4.3. cDNA Probe Synthesis

1. Prepare a master mix for all labeling reactions plus one extra reaction to ensure that you have sufficient volume. Combine the following (per reaction) in a 0.5-mL microcentrifuge tube at room temperature (*see* **Note 8**): 4 µL of 5X BD PowerScript reaction buffer, 2 µL of 10X dNTP mix (for dATP label), 7 µL of [α-^{33}P]dATP (>2500 Ci/mmol, 10 µCi/µL), and 1.5 µL of DTT (100 m*M*), for a total volume of 11.5 µL.
2. Preheat a PCR thermal cycler to 65°C.
3. Add 4 µL of 10X random primer mix to the resuspended beads. Mix well by pipeting.
4. Incubate the beads and primer mix in the preheated thermal cycler at 65°C for 3 min.
5. Reduce the temperature of the thermal cycler to 42°C; incubate the tubes for 2 min. During this incubation, add 2 µL of BD PowerScript Reverse Transcriptase per reaction to the master mix (*see* **Note 8**). Mix by pipeting, and keep the master mix at room temperature.
6. After completion of the 2-min incubation at 42°C, add 13.5 µL of master mix to each reaction tube. Mix the contents of the tubes thoroughly by pipeting, and immediately return them to the thermal cycler.
7. Incubate the tubes at 42°C for 30 min.
8. Add 2 µL of 10X termination mix, and mix well.
9. Dilute the probe synthesis reactions to a 200-µL total volume with buffer NT2; mix well by pipeting. Separate the beads and pipet the supernatant (approx 20 µL) into 180 µL of buffer NT2.
10. Place a NucleoSpin extraction spin column into a 2-mL collection tube, and pipet the sample into the column. Centrifuge at 16,000*g* for 1 min. Discard the collection tube and flowthrough into the appropriate container for radioactive waste.
11. Insert the NucleoSpin column into a fresh 2-mL collection tube. Add 400 µL of buffer NT3 to the column. Centrifuge at 16,000*g* for 1 min. Discard the collection tube and flowthrough.
12. Repeat **step 11** twice.
13. Transfer the NucleoSpin column to a clean 1.5-mL microcentrifuge tube. Add 100 µL of buffer NE, and allow the column to soak for 2 min.
14. Centrifuge at 16,000*g* for 1 min to elute the purified probe.
15. Check the radioactivity of the probe by scintillation counting:
    a. Add 5 µL of each purified probe to 5 mL of scintillation fluid in separate scintillation-counter vials.
    b. Count the samples on the ^{33}P channel. Multiply the counts by a dilution factor of 20. Probes should yield a total of $5–25 \times 10^6$ cpm.

16. Discard the flowthrough fractions, columns, and elution tubes in the appropriate container for radioactive waste.

3.5. Hybridization of cDNA Probes to Plastic Microarray

The hybridization procedure described next is optimized for use with the BD Atlas Plastic Array Hybridization Box. Use of this hybridization box, or an equivalent container, is strongly recommended to ensure that the array is kept flat during the experiment.

- As a general rule for any hybridization vessel, ensure that there is sufficient BD PlasticHyb hybridization solution to completely bathe the microarray.
- Ensure that the printed surface of the microarray is facing up.
- Ensure that the box is well sealed and continuously agitated on a rocking platform during all steps.

1. Fill a hybridization box approx 80% full with H_2O and warm to 55–60°C. Also prewarm 30 mL of BD PlasticHyb hybridization solution at 60°C in a separate container.
2. Carefully place the microarray into the hybridization box containing the prewarmed H_2O, with the printed surface **facing up**.
3. Pour off the H_2O and replace with 15 mL of prewarmed hybridization solution. Firmly attach the lid.
4. Rock the microarray for 10–30 min at 60°C.
5. To prepare your probe for hybridization, incubate the probe in a boiling (95–100°C) water bath for 2 min. Then incubate on ice for 2 min. If desired, Step 6 can be performed as an alternative to heat denaturation.
6. Optional: We find that boiling is adequate to denature probes; however, if you prefer an alkaline denaturing procedure, you may use the following steps instead:
    a. Mix:

labeled probe (entire sample)	approx 100 µL
10X denaturing solution (1 M NaOH, 10 mM EDTA) (or 1/10 Total Volume)	approx 11 µL
Total Volume	approx 111 µL

    b. Incubate at 68°C for 20 min.
    c. Add the following to your denatured probe:

2X neutralizing solution (1 M NaH_2PO_4 [pH 7.0]) (or 1/2 Total Volume)	approx 115 µL
Total Volume	approx 230 µL

    d. Continue incubating at 68°C for 10 min.
7. Combine the denatured probe with 15 mL of prewarmed hybridization solution in a disposable 50-mL or 15-mL plastic tube. Make sure that the two solutions are thoroughly mixed together.

8. Carefully pour off the pre-rinsing solution from the microarray, and add the denatured probe prepared in Step 7 to the microarray. Ensure that the printed surface is facing up. Firmly attach the lid.
9. Hybridize overnight with continuous rocking at 60°C. Ensure that all regions of the plastic are in contact with the hybridization solution at all times. Full coverage is critical because the plastic is nonporous and does not absorb liquid.
10. The next day, prewarm 300 mL of high salt wash solution (2X SSC, 0.1% SDS), and 300 mL of low salt wash solution 1 (0.1X SSC, 0.1% SDS) to 58–60°C. Fill a 500 mL beaker with room-temperature low salt wash solution 2 (0.1X SSC).
11. Open the box containing your hybridizing microarray. Leaving the microarray inside the box, carefully pour off the hybridization solution into an appropriate container for radioactive waste. Immediately fill the box with 40–50 mL prewarmed high salt wash solution. Do not allow the microarray to dry. Do not allow the box or the wash solution to cool. Reattach the lid and rock for 5 min at 58°C to remove residual radioactive probe. Discard the wash solution.
12. Repeat **step 11** one time.
13. After pouring off the second high salt wash, fill the hybridization box with 40–50 mL prewarmed low salt wash solution 1 (from **step 10**). Wash the microarray in a 58°C incubator for 5 min.
14. Repeat **step 13** one time.
15. Reduce the temperature of the hybridization oven to 25–30°C. Pour off the wash solution, and fill the box to approximately 80% capacity (40–50 mL) with **room-temperature** low salt wash solution 2 (0.1X SSC). Rock the microarray for 5 min.
16. Remove the microarray from the hybridization box using forceps. Immediately transfer the microarray to the beaker of room-temperature low salt wash solution 2 (0.1X SSC). Rinse the microarray by dipping it several times into the wash solution.
17. Remove the microarray from the beaker of low salt wash solution very slowly, allowing the wash solution to drain off the surface. Usually, only small droplets of wash solution will remain on the microarray after this step. If large droplets are present, dip the microarray into the low salt bath and slowly remove it again. If large droplets still remain, they can be removed by absorption with a dust-free tissue. Failure to remove these large droplets can cause "plaques" in the microarray image.
18. Allow the microarray to air-dry completely (about 5–10 min). **Do not** use a heating device to rapidly dry the microarray.
19. Expose the printed surface of the microarray to a phosphorimaging screen suitable for ^{33}P detection. If for any reason the plastic is warped, affix the microarray to the phosphorimaging cassette with adhesive tape along the entire length of the microarray edges. Ensure complete contact with the phosphorimager screen. **Do not** cover the microarray with plastic wrap. Typical exposure times range from 12–72 h, but longer exposures can be performed, as necessary. When the exposure is complete, scan the phosphorimager screen at a resolution of ≤50 µm

(*see* **Note 9**). The resolution may need to be optimized for arrays with different printing parameters. Following exposure, proceed directly to stripping or store the microarray at room temperature in the dark.

3.6. Stripping of cDNA Probes from Atlas Plastic Microarray

To reuse a plastic array after phosphorimaging, you may remove the cDNA probe by stripping. For successful stripping, microarrays must be stored at room temperature in the dark at all times following hybridization and analysis. For best results, strip the array as soon as possible after completing exposure.

1. Preheat 40 mL of 0.1 M Na_2CO_3 to 80°C in a BD Atlas Plastic Array Hybridization Box.
2. Insert the microarray into the box with the printed surface facing down. Ensure that no large bubbles are trapped under the array, and attach the lid. Incubate at 80°C for 10 min (or up to 20 min) on a plate vortexor.
3. Remove the microarray from the solution and immediately rinse it in a bath of room temperature deionized H_2O.
4. Allow the microarray to air-dry completely. At this stage, it is not critical to remove all H_2O droplets before drying.
5. Check the efficiency of stripping with a Geiger hand counter and by exposure to a phosphorimaging screen. If radioactivity can still be detected, repeat the stripping procedure (**steps 1–4**). Spots producing very strong signals may still be detectable after stripping (*see* **Note 10**).
6. Store the microarray at room temperature in the dark until needed.

3.7. Interpretation of Results

3.7.1. Sensitivity of Detection and Signal Resolution

After hybridization and washing, we recommend that you perform a "trial run" exposure of the plastic microarrays to a phosphorimaging screen. This will allow you to assess the sensitivity and quality of the hybridization pattern so that you can determine the optimal exposure time for the experiment. As long as the RNA is of high quality, the signals corresponding to medium- to high-abundance mRNAs (0.05–0.5% of poly A^+ RNA) can be easily detected after several hours or an overnight exposure (*see* **Note 11**). Rare- to medium-abundance mRNAs can sometimes be seen after overnight exposures. However, longer exposures may be required. The exact number of hybridization signals observed depends on the complexity of the experimental RNA sample and may differ by severalfold. The practical limit for sensitivity is the level of background generated by nonspecific hybridization of the probe to the plastic support. Longer exposure times (>7 d) are useful only if the background level is low. Some samples may produce signals that are similar or even higher in intensity than the abundant housekeeping genes. After an overnight exposure with ^{33}P-labeled probes, you should observe signals for the most abundant housekeeping genes, including ubiquitin, phospholipase A_2, α-tubulin, β-actin, and G3PDH. These genes are expressed at about 0.1–0.5% abundance in mammalian tissues or cells and can be used as universal positive controls, if included on the array. Note that the ratio of intensities of signals for different housekeeping genes may differ by two- to fivefold for different tissues or cells. After overnight exposure, one generally does not see hybridization with blank regions of the plastic or with any negative controls, if present, on the array. If you observe non-specific background in the form of a sprinkle-like pattern of small dots, *see* **Note 12**. If you observe hybridization for the negative controls, *see* **Note 13** for recommendations on reducing nonspecific hybridization.

Another important factor in signal detection is signal resolution. If you obtain "fuzzy" hybridization signals, the problem may be that the microarray was not in complete, even contact with the phosphorimager screen (*see* **Note 14**). Poor signal resolution may also be owing to irregular spot morphology. *See* **Note 15** for possible causes and solutions.

3.7.2. Normalization of Hybridization Signals

The best approach for comparing hybridization signals for different samples is to equalize the intensity of the hybridization signals by adjusting exposure times. If one array is uniformly darker than the other, adjust the exposure time of one array until the overall signal is approximately the same on both arrays. The most common reason for different overall hybridization intensities is the quality of RNA samples used to prepare the hybridization probes. For more details about RNA quality, refer to **Notes 4**, **5**, and **7**. In our experience, it is most effective and convenient to normalize arrays based on the overall signal from all genes on the array.

As an alternative to normalization based on the overall level of signal, some researchers prefer to identify one or more housekeeping genes that generate equally intense hybridization signals for both samples being compared. This housekeeping gene or genes can then serve as a normalization standard. In cells or tissues that are closely related—i.e., where only a few genes change their expression levels—the expression of housekeeping genes generally remains constant. However, the expression levels of individual housekeeping genes may vary depending on your experimental system, especially if different tissues are being compared.

4. Notes

1. Affixing a small piece of adhesive tape to the corner of the protective sheet as a "handle" may make it easier to peel the sheet back from the surface of the film.
2. If using a tray, a MicroSeal A Film with both protective sheets removed can also be used to create an adhesive "pad" under the Atlas Plastic Film (*see* **Subheading 2.1.2.**).

3. For very RNase-rich samples (e.g., pancreas, liver, spleen), we recommend that you perform a third or fourth round of phenol:chloroform extraction.
4. Impurities in RNA samples can inhibit RT. In this case, you may need to perform additional steps to purify the total RNA starting material. Try treating the total RNA twice with phenol:chloroform and once with chloroform, followed by precipitation with 1/10 vol of 2 M NaOAc (pH 4.5) and 2.5 vol of ethanol. This will help ensure the removal of any protein and other impurities that may not have been removed effectively during initial RNA purification. You can perform additional RNA purification methods such as CsCl centrifugation or gel filtration, if necessary. Such procedures should be optimized for each particular tissue/cell type separately.
5. If, on a denaturing formaldehyde/agarose/EtBr gel, the total RNA appears as a smear that is no larger than 2 kb, the RNA may be degraded. If this is the case, we suggest that you prepare fresh RNA after checking the purification reagents for RNase or other impurities. If problems persist, you may need to find another source of tissue/cells.
6. Be sure to work through the enrichment/probe synthesis steps quickly, without pausing. Additionally, to help reduce any chance of RNA degradation, you may add 100 U of Ambion's ANTI-RNase (cat. no. 2692) after adding magnetic beads to the sample.
7. For synthesizing probes with the highest sensitivity, we recommend using as much RNA as possible, up to the 50-µg limit.
8. If desired, you may use the wild-type MMLV RT provided with the BD Atlas Pure Kit instead; however, you should use the same enzyme to label all probes that will be directly compared. Ensure that you use the correct 5X reaction buffer.
9. We routinely use a phosphorimager at 25–50 µm resolution. However, the required resolution may vary depending on the specific printing parameters.
10. If you cannot successfully strip probe from the microarray, use the following guidelines to troubleshoot:
    a. *Microarray stored incorrectly:* Avoid exposure to sunlight and do not expose to temperatures higher than room temperature.
    b. *Microarray inappropriately heated:* Do not use a heating device to dry the plastic surface after hybridization.
    c. *Old or incorrect stripping solution:* Make fresh stripping solution ensuring the correct composition as described in **Subheading 2.3.**
    d. *Microarray not fully bathed during stripping:* Ensure that the microarray is covered at all times during stripping.
    e. Plastic microarrays should be stripped as soon as possible. After several weeks of storage, stripping may be difficult.
11. Because of sequence-dependent hybridization characteristics and variations inherent in any hybridization reaction, array data should be considered semiquantitative. We strongly recommend that you corroborate the results of your experiment using RT-PCR.

Plastic Microarrays

12. If you observe non-specific background in the form of a sprinkle-like pattern of small dots, your RNA may not be of sufficient quality. Sometimes this background pattern is misinterpreted as a printing error. To eliminate this problem, we strongly recommend using the spin columns provided in the NucleoSpin® RNA II Kit (cat. no. K3064-1) to purify your RNA, even if the sample was isolated using another method.
13. Several factors may account for high levels of nonspecific background:
    a. The printed microarray surface did not remain adequately covered with hybridization or wash solution.
    b. The microarray may have partially dried before washing was completed. Work swiftly to ensure that the microarray does not dry between washing steps.
    c. The radiolabeled nucleotide was too old. If the ^{33}P-dATP you used for labeling was older than 2 wk, make a new probe with fresh nucleotide. Some lots of dATP may inhibit incorporation.
    d. The probe was not purified correctly. If, for some reason, you did not purify the cDNA probe, use the NucleoSpin extraction spin columns. Do not use another purification method.
    e. The microarray was stripped and reprobed too many times. Generally, plastic microarrays can be stripped and reprobed at least three times.
14. If you obtain "fuzzy" hybridization signals, the problem may be that the microarray was not in complete, even contact with the phosphorimager screen. Use the following guidelines to ensure that the microarray is in close contact to the screen:
    a. Attach the microarray to the phosphorimager cassette using adhesive tape along the entire length of the microarray edges.
    b. Place a cardboard insert in the phosphorimaging cassette to bring the screen flush with the microarray.
15. Although BD Atlas Plastic Films display excellent printability, certain problems in the printing process can result in irregular spot morphology. For best results, use the BD Atlas Plastic Printing Buffer included with BD Atlas Plastic Films; do not substitute your own reagents. If you do encounter problems, use the following guidelines to troubleshoot:
    a. *Bent, broken, or dirty pins:* Thoroughly clean the pin tool between printings. If problems persist, examine the pin tool and scan for damaged pins. Replace any suspect pins.
    b. *Contaminants in printing solution:* Impurities such as salts, detergents, and polysaccharides in the nucleic acids used for printing can affect spot morphology. Use only highly purified nucleic acids resuspended or eluted in nuclease-free H_2O.
    c. *Printing process errors:* Errors in printing, such as an abbreviated print time or excess speed in pin retraction, can lead to donut-shaped spots. These spots are formed when the printing solution is deposited unevenly, resulting in a ring of printed material at the outer edge of the spot surrounding a nearly empty center. Be sure to follow the manufacturer's printing guidelines for your arraying device.

Reference

1. Duggan, D. J., Bittner, M., Chen, Y., Meltzer, P., and Trent, J. M. (1999) Expression profiling using cDNA microarrays. *Nat. Genet.* **21,** 10–4.

Suggested Reading

Cheung, V. G., Morley, M., Aguilar, F., Massimi, A., Kucherlapati, R., and Childs, G. (1999) Making and reading microarrays. *Nat. Genet.* **21,** 15–19.

DeRisi, J. L., Iyer, V. R., and Brown, P. O. (1997) Exploring the metabolic and genetic control of gene expression on a genomic scale. *Science* **278,** 680–686.

DeRisi, J., Penland, L., Brown, P. O., Bittner, M. L., Meltzer, P. S., Ray, M., Chen, Y., Su, Y. A., and Trent, J. M. (1996) Use of a cDNA microarray to analyse gene expression patterns in human cancer. *Nat. Genet.* **14,** 457–460.

Eisen, M. B. and Brown, P. O. (1999) DNA arrays for analysis of gene expression. *Methods Enzymol.* **303,** 179–205.

Lockhart, D. J. and Winzeler, E. A. (2000) Genomics, gene expression and DNA arrays. *Nature* **405,** 827–836.

4

Preparing Fluorescent Probes for Microarray Studies

Charlie C. Xiang and Michael J. Brownstein

1. Introduction

A number of articles have been published describing methods to produce fluorescent probes from RNA (or DNA) samples. These methods are conceptually similar. Broadly speaking, they involve some or all of the following procedures: template amplification, template transcription with concomitant incorporation of modified bases into the transcribed products (DNA or RNA), and signal amplification. The simplest technique relies on the direct incorporation of Cy3- or Cy5-labeled nucleotides into oligo-dT-primed cDNA by reverse transcriptase (RT). This method has proven to be reasonably robust, but it requires 50–200 µg of total RNA or 2–5 µg of poly(A) RNA per labeling *(1–4)*. Thus, rather large tissue samples are required for each study.

A variation on the aforementioned technique involves using aminoallyl-modified bases *(5,6)* instead of dye-labeled ones. After the DNA is produced, dyes are coupled to the free amines. This indirect labeling procedure is no less demanding of RNA than the direct labeling method, but it is somewhat less expensive to use. Furthermore, it has been argued that incorporation of dye-labeled bases into DNA is somewhat biased, and the use of a single, aminoallyl-modified base circumvents this problem; however, we have not found that direct labeling results in dye bias.

Researchers have attempted to improve on the methods described so that they can study smaller RNA samples. One obvious modification was to use random instead of oligo(dT) priming *(7)*. This permits one to generate several probes from each transcript and to make more DNA per microgram of starting material. When we tried this, the result was encouraging, but not as good as we had hoped it would be. Therefore, in the method described herein, we made one further modification: we added an amine-modified base to the ends of the primers so that we could introduce dye into each cDNA and onto the end of each product as well *(8)*. This allowed us to label 1 μg of total RNA in such a way that reproducible, quantitative experiments could be performed.

In principle, the technique that we developed can be used in tandem with signal amplification methods including dendrimer labeling *(9)* or tyramide amplification *(10)*. This would have to be done with care to avoid signal saturation.

Another way to reduce our method's RNA requirement would be to amplify the template before making labeled probe. Two methods to accomplish this are published in the literature, the first based on production of antisense RNA (aRNA) by T7 polymerase *(11,12)*, the second based on RT-polymerase chain reaction (PCR) *(13)*. Neither method produces a faithful replica of the starting pool of transcripts, but the former has been used successfully in many published studies.

Finally, it is worth mentioning the technique of choice for Affymetrix chip arrays, which combines template and signal amplification. In this instance, double-stranded cDNA, which has a T7 polymerase recognition sequence on the 3′ end, is prepared. Biotin-labeled cRNA made from this template is hybridized to arrays, and then these are stained with streptavidin-phycoerythrin *(14)*.

2. Materials

1. TRIzol reagent (Invitrogen/Life Technologies, Carlsbad, CA).
2. Phase Lock Gel (Eppendorf).
3. Absolutely RNA microprep kit (Stratagene, La Jolla, CA).
4. RNasin RNase inhibitor (Promega, Madison, WI).
5. Superscript II RT, 5X first-strand buffer, 0.1 M dithiothreitol (DTT) (Invitrogen/Life Technologies).
6. Microcon 30 concentrator (Millipore, Bedford, MA).
7. Monofunctional NHS-ester Cy3 and Cy5 dyes and dNTPs (Amersham Pharmacia, Piscataway, NJ).
8. QIAquick PCR purification kit and MinElute PCR purification kit (Qiagen, Valencia, CA).
9. 5-[3-Aminoallyl]-2-deoxyuridine 5-triphosphate (aa-dUTP) (Sigma, St. Louis, MO).
10. Oligo dT and random hexamer primers (Invitrogen/Life Technologies). We purchased custom-synthesized amine-modified random primers from Sigma Genosys (Woodlands, TX). These primers are random hexamers with aminoC6-dTTP added to the 5′ end.

3. Methods (see Notes 1–3)
3.1. Isolation of RNA

We have routinely extracted total RNAs using TRIzol reagent (Invitrogen/Life Technologies) according to the manufacturer's instructions. It may be helpful to centrifuge the crude homogenate at 12,000g at 4°C for 10 min to pellet polysaccharides and DNA, and to precipitate the RNA by adding 0.25 mL of isopropanol and 0.25 mL of 0.8 M sodium citrate and 1.2 M NaCl per milliliter of TRIzol reagent. This keeps proteoglycans and polysaccharides in a soluble form and yields a cleaner product when tissues rich in these substances, such as liver, are extracted. In addition, using Phase Lock Gel (Eppendorf) eliminates interphase-protein contamination and improves recoveries. Finally, we have used commercially available kits for RNA purification, among them the Absolutely RNA microprep kit (Stratagene), which works well for isolating RNA from small tissue samples in good yield.

3.2. cDNA Synthesis.

1. Combine 0.1–5 µg of total RNA (15.5 µL) with amine-modified random primer (2 µg/µL, 2 µL) and RNase inhibitor (5 U/µL, 1 µL).
2. Incubate the mixture at 70°C for 10 min, and chill on ice for 10 min.
3. Add the primer/RNA solution to the RT mix (6 µL of 5X first-strand buffer; 50X aa-dUTP/dNTPs [25 mM dATP, dGTP, and dCTP; 15 mM dTTP; and 10 mM aminoallyl dUTP], 3 µL of 0.1 M DTT, 2 µL of SSII RT) and incubate at 42°C for 2 h.

3.3. RNA Removal

1. Terminate the reaction by adding EDTA (0.5 M, 10 µL), and hydrolyze the RNA with NaOH (1 M, 10 µL) at 65°C for 30 min.
2. Neutralize the solution with HCl (1 M, 10 µL).

3.4. Purification of cDNA

Use MinElute PCR purification kits to purify the cDNA.

1. Fill a microcentrifuge tube with 300 µL of buffer PB. Add 60 µL of the neutralized reaction solution to buffer PB.
2. Place a MinElute column in a 2-mL collection tube in a suitable rack.
3. To bind DNA, apply the sample to the MinElute column and centrifuge for 1 min. For maximum recovery, transfer all traces of sample to the column.
4. Pour the flowthrough back into the column and centrifuge again for 1 min. Discard the flowthrough.
5. Place the MinElute column back into the original collection tube.

6. Add 750 µL of buffer PE to the MinElute column, incubate for 5 min at room temperature, and centrifuge for 1 min.
7. Discard the flowthrough and place the MinElute column back in the same tube.
8. Centrifuge the column for an additional 1 min at maximum speed.
9. Place the MinElute column in a clean 1.5-mL microcentrifuge tube.
10. To elute the DNA, add 10 µL of H_2O (pH between 7.0 and 8.5) to the center of the membrane, let the column stand for 1 min, and then centrifuge for 5 min. The average eluate is 9 µL out of the 10 µL applied.
11. Elute the DNA twice more with 10 µL of H_2O, collecting a total of 27 µL of purified cDNA.

3.5. Dye Coupling

1. Add 3 µL of 1 M sodium bicarbonate, pH 9.3, to the cDNA solution.
2. Add 1 µL of dye solution (NHS-ester Cy3 or Cy5, 62.5 µg/µL in dimethylsulfoxide) to the cDNA solution, and mix by pipeting the resulting solution up and down several times; wrap the tube in aluminium foil, and incubate at room temperature for 1 h in an orbital shaker (USA Scientific, Ocala, FL).
3. Stop the labeling reaction with 4.5 µL of 4 M hydroxylamine hydrochloride. Mix and briefly centrifuge.
4. Incubate the tube for 30 min at room temperature in the dark.

3.6. Purification of Dye-Labeled Probes

Clean the probes with a QIAquick PCR purification kit.

1. Combine the Cy3- and Cy5-labeled products; add 30 µL of water and then 500 µL of Buffer PB.
2. Apply the solution to a QIAquick column and spin at 16,000g for 1 min. Discard the flowthrough.
3. To wash the column, add 750 µL of buffer PE and spin for 1 min. Discard the flowthrough.
4. Repeat the wash step.
5. Spin the column once again to remove residual ethanol.
6. Place a fresh collection tube beneath the column. Add 30 µL of buffer EB to the column and incubate for 1 min at room temperature.
7. Spin at 16,000g for 1 min and repeat the elution step once.
8. Partially dry the eluate in a vacuum centrifuge and adjust the volume to 23 µL with water.

3.7. Hybridization and Wash Conditions

Plan to use probes soon after they have been prepared; they are not stable for long periods. Similarly, plan to read the arrays as quickly as possible. We read arrays with a GenePix 4000A scanner (Axon, Foster City, CA) at 10-μM resolution and variable photomultiplier tube voltage setting to obtain the maximal signal intensities with <1% probe saturation, and we analyze the resulting images with IPLab (Fairfax, VA) and ArraySuite (NHGRI, Bethesda, MD) software. We recommend using the method of Chen et al. *(15)* to filter bad data points.

1. Add 4.5 µL of 20X saline sodium citrate (SSC), 2 µL of poly(A) (10 mg/mL), and 0.6 µL of 10% sodium dodecyl sulfate (SDS), and denature at 100°C for 2 min.
2. Pipet the solution onto the array, apply a cover slip, and place the slide in a hybridization chamber (Corning, Corning, NY).
3. Incubate the array in a 65°C water bath for 16–24 h.
4. Wash the slide with 0.5X SSC, 0.01% SDS followed by 0.06X SSC at room temperature, 10 min each.
5. Place the slide in a 50-mL Falcon tube and spin for 5 min at 800 rpm at room temperature.

4. Notes

1. The method described above has proven to be quite reliable. It gives results that are comparable with those obtained with the conventional technique of oligo(dT) priming and direct incorporation of dye-labeled bases. In fact, using the new method, we see an improved signal-to-noise ratio and less variability.
2. At the expense of some deterioration in these parameters, it should be possible to reduce dramatically the amount of input RNA required for an array experiment by template and/or signal amplification.
3. We have not used the method described with spotted oligonucleotide arrays but would expect the signals to be somewhat weaker than those observed using cDNA arrays.

References

1. Duggan, D. J., Bittner, M., Chen, Y., Meltzer, P., and Trent, J. M. (1999) Expression profiling using cDNA microarrays. *Nat. Genet.* **21,** 10–14.
2. National Human Genome Research Institute. Division of Intramural Research Microarray Project (uAP). www.nhgri.nih.gov/DIR/Microarray/main.html.
3. Schena, M., Shalon, D., Davis, R. W., and Brown, P. O. (1995) Quantitative monitoring of gene expression patterns with a complementary DNA microarray. *Science* **270,** 467–470.
4. Stanford University. School of Medicine. The Brown Lab. *The MGuide.* http://cmgm.stanford.edu/pbrown/mguide/index.html.
5. Schroeder, B. G., Peterson, L. M., and Fleischmann, R. D. (2002) Improving quantitation and reproducibility in Mycobacterium tuberculosis DNA microarrays. *J. Mol. Microbiol. Biotechnol.* **4,** 123–126.
6. Xiang, C. C., Kozhich, O., Chen, M., Inman, J. M., Phan, Q. N., Chen, Y., and Brownstein, M. J. (2002) *Nat. Biotechnol.* **20,** 738–742.

7. Yu, J., Othman, M. I., Farjo, R., Zareparsi, S., MacMee, S. P., Yoshida, S., and Swaroop, A. (2002) Evaluation and optimization of procedures for target labeling and hybridization of cDNA microarrays. *Mol. Vis.* **8,** 130–137.
8. Yue, H., Eastman, P. S., Wang, B. B., Minor, J., Doctolero, M. H., Nuttall, R. L., Stack, R., Becker, J. W., Montgovery, J. R., Vainer, M., and Johnston, R. (2001) An evaluation of the performance of cDNA microarrays for detecting changes in global mRNA expression. *Nucleic Acids Res.* **29,** e41.
9. Stears, R. L., Getts, R. C., and Gullans, S. R. (2000) A novel, sensitive detection system for high-density microarrays using dendrimer technology. *Physiol. Genomics* **3,** 93–99.
10. Karsten, S. L., Van Deerlin, V. M. D., Sabatti, C., Gill, L. H., and Geschwind, D. H. (2002) An evaluation of tyramide signal amplification and archived fixed and frozen tissue in microarray gene expression analysis. *Nucleic Acids Res.* **30,** e4.
11. Van Gelder, R. N., von Zastrow, M. E., Yool, A., Dement, W. C., Barchas, J. D., and Eberwine, J. H. (1990) Amplified RNA synthesized from limited quantities of heterogeneous cDNA. *Proc. Natl. Acad. Sci. USA* **87,** 1663–1667.
12. Wang, E., Miller, L. D., Ohnmacht, G. A., Liu, E. T., and Marincola, F. M. (2000) *Nat. Biotechnol.* **18,** 457–459.
13. Hughes, T. R., Mao, M., Jones, A. R., et al. (2001) *Nat. Biotechnol.* **19,** 342–347.
14. Mahadevappa, M. and Warrington, J. A. (1999) A high-density probe array sample preparation method using 10- to 100-fold fewer cells. *Nat. Biotechnol.* **17,** 1134–1136.
15. Chen, Y., Kamat, V., Dougherty, E. R., Bittner, M. L., Meltzer, P. S., and Trent, J. M. Ratio statistics of gene expression levels and applications to microarray data analysis. *Bioinformatics,* in press.

5

Escherichia coli Spotted Double-Strand DNA Microarrays

RNA Extraction, Labeling, Hybridization, Quality Control, and Data Management

Arkady B. Khodursky, Jonathan A. Bernstein, Brian J. Peter, Virgil Rhodius, Volker F. Wendisch, and Daniel P. Zimmer

1. Introduction

Highly parallel hybridization of nucleic acids on glass slides has successfully been applied to measure RNA and DNA abundances in *Escherichia coli* (*1–4*). In this chapter, we summarize our experience in working with *E. coli* DNA microarrays accumulated over a 4-yr period. Typically, we printed and used *E. coli* DNA microarrays containing roughly 6000 spotted elements. These included 4200 amplicons of *E. coli* open reading frames (ORFs), 112 amplicons of genes encoding stable RNAs, and more than 1500 control elements and replicates. We describe the methods for total RNA extraction, mRNA enrichment of total RNA, cDNA labeling via direct and indirect incorporation of fluorophors, and microarray hybridization. Additionally, we present strategies for optimizing microarray hybridizations and descriptions of several Internet-based tools useful in analyzing data from array experiments.

2. Materials

2.1. Isolation of RNA (see Note 1)

2.1.1. A. Total RNA Isolation

1. TE buffer, pH 8.0 (cat. no. 9849; Ambion).
2. 10% (w/v) sodium dodecyl sulfate (SDS) (cat. no. 9822; Ambion). Also prepare a 5% solution.

3. Ultrapure phenol (redistilled, crystalline) (cat. no. 15509; Invitrogen/Gibco-BRL). Note that water-saturated or acid phenol should be used for RNA extractions. DNA is partially hydrolyzed at acid pH *(5,6)*.
4. RNeasy (mini, midi, or maxi) Kit (cat. no. 74104, 75144, or 75162; Qiagen).
5. RNase-free DNase Set (cat. no. 79254; Qiagen).

2.1.2. B. mRNA Enrichment

1. Zirconium/silica beads (cat. no. 11079101Z; Biospec, Bartlesville, OK).
2. Mini-BeadBeater (cat. no. 3110BX; Biospec).
3. Tri Reagent LS (cat. no. TS 120; Molecular Research Center, Cincinnati, OH).
4. Polyadenylation buffer: 40 mM Tris, pH 8.0; 250 mM NaCl, 10 mM MgCl$_2$, 50 µg/mL of bovine serum albumin (BSA), 5 mM MnCl$_2$, 0.8 U/µL of RNase inhibitor, 0.01 U/µL of RNase-free DNase I, and 400 µM adenosine triphosphatase (ATP). Always prepare the polyadenylation buffer freshly. A 5X stock solution containing 200 mM Tris, pH 8.0, 1.25 M NaCl, 50 mM MgCl$_2$, and 250 µg/mL of BSA may be prepared and stored at –20°C. Also, prepare and freeze 10X stock solutions of 50 mM MnCl$_2$ and of 4 mM ATP.
5. *E. coli* poly(A) polymerase (cat. no. E2180Y; Amersham Pharmacia Biotech).
6. Oligotex mRNA purification system (cat. no. 70022; Qiagen).

2.2. DNase I Treatment of RNA Samples

1. RNase inhibitor (40 U/µL) (cat. no. 799017; Boehringer Mannheim).
2. RNase-free DNase I (10 U/µL) (cat. no. 776785; Boehringer Mannheim).
3. 5X DNase I buffer: 50 mM MgCl$_2$, 50 mM Tris-HCl (pH 7.5), 5 mM EDTA, 5 mM dithiothreitol (DTT).
4. Ethanol (absolute, denatured [SDA Formula 3A], ACS) from VWR (case of 12 × 500 mL; cat. no. MK701904).

2.3. Denaturing Formaldehyde Gel

1. Agarose.
2. Formaldehyde gel running buffer (20 mM MOPS, 5 mM sodium acetate, 1 mM EDTA, 0.74% formaldehyde, pH 7.0) The formaldehyde buffer is made up from a 10X stock solution. (One liter of 10X formaldehyde gel buffer is 41.9 g of MOPS [free acid], 6.8 g of Na acetate•3H$_2$O, 20 mL of 0.5 M EDTA, pH to 7.0 with NaOH, 20 mL of 37% [12.3 M] formaldehyde, and diethylpyrocarbonate [DEPC]-treated H$_2$O to 1 L.)
3. 5X RNA loading buffer. (Ten milliliters of loading buffer is 16 µL of saturated aqueous bromophenol blue solution, 80 µL of 500 mM EDTA, pH 8.0, 720 µL of 37% [12.3 M] formaldehyde, 2 mL of glycerol, 3.1 mL of formamide, 4 mL of 10X formaldehyde gel buffer, and DEPC-treated H$_2$O to 10 mL.)

2.4. cDNA Synthesis

1. Random hexamer primers p(dN$_6$) (cat. no. 27-2166; Amersham Pharmacia Biotech).
2. Millipore Microcon YM-30 filters (cat. no. 42410; distributed by Fisher).

E. coli Spotted Double-Strand DNA Microarrays

2.4.1. Indirect Labeling

1. StrataScript RNase H⁻ Reverse Transcriptase (RT) and 10X StrataScript RT Buffer (cat. no. 600085; Stratagene).
2. RNase Inhibitor (40 U/µL) (cat. no. 300151; Stratagene).
3. Aminoallyl dUTP (sodium salt) (cat. no. A 0410; Sigma, St. Louis, MO).
4. 50X Aminoallyl dUTP/dNTP labeling mix (25 mM each of dATP, dCTP, and dGTP, 15 mM aminoallyl dUTP, and 10 mM dTTP). All four dNTPs are sold as a set (cat. no. 1 969 064; Roche). Store the 50X mix as small aliquots at –20°C.
5. 0.1 M stock of DTT; prepare and store at –20°C.
6. Freshly prepared 1 N NaOH.
7. Cy3 and Cy5 monofunctional dye (Cy3 monoreactive dye pack, cat. no. PA23001; Cy5 monoreactive dye pack, cat. no. PA25001; Amersham Pharmacia Biotech). Each pack contains five vials sealed in foil bags, each with a desiccant capsule. Note that the dyes are light sensitive and extremely unstable in aqueous solution. When ready for use, thoroughly resuspend an entire vial in 10–10.5 µL of dimethylulfoxide (DMSO). Use the required amount for the current experiment and then divide the remainder into 1.25-µL aliquots in 0.5-mL microfuge tubes. Dry the dyes immediately in a SpeedVac in the dark to minimize light exposure (approx 60 min). Seal the tubes with parafilm, place in the original foil bags containing the blue desiccant capsule, seal, and store in the dark at 4°C for up to 3 wk. When ready to use a dried aliquot, resuspend in 1.25 µL of DMSO immediately prior to use.
8. QIAquick PCR Purification Kit (50) (cat. no. 28104; Qiagen).

2.4.2. Direct Labeling

1. SuperScript II RNase H⁻ RT (cat. no. 18064-014; Gibco-BRL), shipped with 5X first-strand buffer and 0.1 M DTT.
2. Ultrapure dNTP set at 100 mM each (cat. no. 27-2035-01; Amersham Pharmacia Biotech). Prepare a 10X dNTP mix (1 mM dATP, 1 mM dGTP, 1 mM dCTP, 0.4 mM TTP).
3. FluoroLink Cy3-dUTP and Cy5-dUTP (cat. no. PA 53022 and 55022; Amersham Pharmacia Biotech).

2.5. Hybridization

1. 20X saline sodium citrate (SSC) (3 M NaCl, 0.3 M sodium citrate, nuclease free) (cat. no. 9764; Ambion). Also prepare a 3X SSC solution.
2. Poly [d(I-C)] (cat. no. 1219847; Roche) or yeast tRNA (cat. no. 109495; Roche). Prepare a 10 mg/mL solution in sterile H_2O and store aliquots at –20°C
3. Millipore UltraFree-MC filters (cat. no. UFC30HVNB; distributed by Fisher).
4. No. 1 Corning cover slips (22 × 22 mm) (cat. no. 12-524C; Fisher) or LifterSlips from Erie Scientific (cat. no. 22x25I-2-4635).
5. Hybridization chambers with screws from Monterey Industries (voice mail: 510-233-3723).

6. Dish-staining complete assembly/glass chambers for post-hybridization washes from ThermoShandon (cat. no. 121).

3. Methods
3.1. Isolation of RNA

Direct or indirect labeling of cDNA synthesized using total RNA as template by the random primer method requires 15–25 µg of DNA-free RNA. The amount of bacterial culture that must be harvested to prepare this quantity of RNA varies depending on the growth medium, strain, cell density, and growth phase. Experience indicates that 1 mL of strain MG1655 taken from an exponential culture in Luria-Bertani (LB) medium at OD_{600} ~0.3–0.5 will yield up to 35 µg of total RNA. However, 10–15 mL of stationary-phase LB medium culture is required to achieve a comparable yield. Two protocols are presented for the preparation of RNA. The first method isolates total RNA, whereas the second is designed to produce samples of enriched mRNA.

3.1.1. Isolation of Total RNA

3.1.1.1. HARVESTING OF CULTURE

1. Grow overnight cultures of the strains from which you intend to harvest RNA.
2. Dilute the overnight cultures according to the experimental design into 100 mL of appropriate media and grow to appropriate density.
3. Harvest the cells by transferring 10 mL of culture to a 15-mL conical tube containing 1.66 mL (1/10 of the final volume) of ice-cold ethanol/phenol stop solution (5% water-saturated phenol [pH <7.0] in ethanol). The stop solution inhibits the degradation of mRNA.
4. Collect the cells by centrifuging at 5900*g* for 5 min at 4°C in a Beckman JA-20 rotor.
5. Pour off the supernatant and remove the remaining media with a pipet (*see* **Note 2**).

3.1.1.2. CELL LYSIS

1. Lyse the cells by resuspending the pellet in a fresh solution of 800 µL of TE, pH 8.0, containing 0.5–5 mg/mL of lysozyme (depending on the strain).
2. Transfer the lysate to RNase-free 2-mL microfuge tubes containing 80 µL of 10% SDS, mix, and place in a water bath at 64°C for 1 to 2 min. The samples should be clear.
5. After incubation, add 88 µL of 1 *M* NaOAc, pH 5.2, (DEPC-treated) and mix by inverting the tubes.

3.1.1.3. HOT PHENOL EXTRACTION

1. Add to the samples an equal volume (1 mL) of water-saturated phenol (pH <7.0).

E. coli Spotted Double-Strand DNA Microarrays

2. Invert 10 times and incubate at 64°C for 6 min. Keep inverting the samples 6–10 times every 40 s or so during the incubation, and then place the tubes on ice to chill for 1 to 2 min.
3. Centrifuge the samples at maximum speed (14,000g) in a benchtop microfuge for 10 min at 4°C.

3.1.1.4. Chloroform Extraction

1. Transfer the aqueous layer to a fresh 2-mL microfuge tube containing an equal volume of chloroform.
2. Invert the tubes 6–10 times, and then centrifuge at maximum speed (14,000g) for 5 min at 4°C.

3.1.1.5. Ethanol Precipitation

1. For each sample, transfer and divide the aqueous layer equally into two 1.5-mL microfuge tubes.
2. Add 1/10 vol of a solution consisting of 3 M Na acetate, pH 5.2, and 1 mM EDTA. Then add 2.5 vol of cold 100% EtOH.
3. Incubate at –80°C for 20 min, and centrifuge at maximum speed (14,000g) in a benchtop microfuge for 25 min at 4°C.
4. Wash the pellet with 1 mL of 70% cold ethanol (made with DEPC-treated water), and centrifuge at maximum speed for 5 min at 4°C.
5. Carefully decant the ethanol and air-dry the pellet (15–20 min in a fume hood).
6. Resuspend each pellet in 100 µL of RNase-free H_2O (DEPC treated), and pool each pair of tubes to give a total volume of 200 µL for each RNA sample.

3.1.2. Isolation and Enrichment of mRNA

3.1.2.1. Harvesting of Culture

1. If the cell extract is to be used in buffer-sensitive downstream applications such as mRNA enrichment by polyadenylation of total RNA (*see* **Note 3**), harvest exponentially growing *E. coli* cells by adding 35 mL of culture to 15 g of ice that has been kept at –80°C.
2. Centrifuge the cells at 7000g for 2 min (or 3500g for 5 min) at 4°C.
3. Aspirate off the media and freeze the pellet by immersing the centrifuge tube in liquid nitrogen. At this point the cells can be stored at –80°C for at least 6 mo. (Liquid nitrogen freezing is also recommended when processing multiple samples for the purpose of treatment synchronization in the RNA extraction procedures. Sample should be thawed on ice before lysis.)

3.1.2.2. Cell Lysis

1. To lyse cells for subsequent mRNA enrichment, resuspend the cell pellet in 1.25 mL of ice-cold polyadenylation buffer.
2. Transfer to a tube containing 0.5 g of 0.1-mm zirconium/silica beads and mechanically lyse the cells for 2 min at 600g in a Mini-BeadBeater (*see* **Note 4**).

3.1.2.3. ENRICHMENT FOR mRNA

Enrichment for mRNA makes use of differentially accessible 3'-termini of mRNA vs rRNA and tRNA in crude extracts *(3)*. In mechanically obtained crude cellular extracts (*see* **Subheading 3.1.2.2.**), the 3'-termini of rRNAs are blocked owing to shielding by ribosomal proteins in (sub)ribosomal complexes and those of tRNAs owing to aminoacylation. By contrast, mRNAs have freely accessible 3'-termini regardless of whether they are free or polysomal.

1. Transfer the cell lysate without zirconium/silica beads to a tube containing 50 U of *E. coli* poly(A) polymerase I *(7)* and incubate for 5 min at 37°C.
2. Prepare RNA from the poly(A) polymerase reaction products using Tri-Reagent-LS according to the manufacturer's instructions.
3. Purify polyadenylated RNA using the Oligotex mRNA purification system according to the manufacturer's instructions. Proceed with DNase treatment in **Subheading 3.2.** The procedure will typically yield ~100 µg of total RNA after cell lysis and ~0.3 µg of enriched mRNA (*see* **Note 5**).

3.2. DNase I Treatment of RNA Samples

It is essential to remove all DNA before the RT reaction. Treat each 200-µL RNA sample with 0.5 µL (20 U) of RNase inhibitor, 50 µL of 5X DNase I buffer, and 1 µL of RNase-free DNase I. Incubate the reactions at 37°C for 30 min.

3.2.1. Phenol and Chloroform Extractions

1. For each extraction, add an equal volume of phenol and/or chloroform, mix by inverting the tubes 6–10 times, centrifuge at maximum speed (14,000*g*) in a microfuge for 2 to 3 min, and transfer the aqueous layer to a fresh microfuge tube.
2. Perform one phenol extraction (use water-saturated phenol, pH <7.0), one phenol (water-saturated)/chloroform (50:50) extraction, and two chloroform extractions.

3.2.2. Ethanol Precipitation

1. For each sample, add 1/10 vol of 3 *M* NaOAc, pH 5.2 (DEPC-treated), and 2.5 vol of cold 100% EtOH.
2. Incubate at −80°C for 20 min and then centrifuge at maximum speed (14,000*g*) in a microfuge for 25 min at 4°C.
3. Carefully remove the ethanol. At this stage the pellet will be barely visible.
4. Wash the pellet with 1 mL of 70% cold ethanol (made with DEPC-treated water) and centrifuge at maximum speed for 5 min at 4°C.
5. Carefully decant the ethanol and air-dry the pellet (15–20 min in the fume hood).
6. Resuspend the pellet in 50 µL of RNase-free H_2O (DEPC treated).

3.3. Determination of RNA Sample Quality

For quantitation, measure the UV absorbance at 260 nm to determine the RNA concentration and A_{260}/A_{280} ratio to determine the RNA purity as described in standard manuals (*see*, e.g., **ref. 8**).

The integrity of the RNA can be analyzed on a denaturing formaldehyde 1% agarose gel. Load 2 µL of each RNA sample. On visualization of the gel, the 16 and 23S RNA should be easily observed. The 23S rRNA should be twice as intense as the 16S rRNA species. Significant downward smearing indicates sample degradation during preparation. Degraded RNA samples should not be used as a template for cDNA synthesis.

3.4. cDNA Synthesis

Following RNA isolation, it is often necessary to concentrate the material either by precipitation or by evaporation through centrifugation under vacuum. If precipitation is performed, it is important to wash the pellet with cold 70% ethanol to remove excess salt prior to labeling with RT. Optimized conditions for indirect (post–cDNA synthesis incorporation via aminoallyl coupling) and direct (incorporation of fluorophors in the course of the reverse transcription reaction) labeling are presented next (*see* **Note 6**).

3.4.1. Indirect Labeling (see **Note 7**)

3.4.1.1. ANNEALING

1. In microfuge tubes mix 16–25 µg of total RNA with 10 µg of random hexamers and DEPC-treated H_2O to a final volume of 20 µL.
2. Incubate the mixture at 70°C for 10 min and then chill on ice for up to 10 min.

3.4.1.2. cDNA SYNTHESIS REACTION (SEE NOTE 6)

1. Add to each annealed RNA/random hexamer mixture 3 µL of 10X StrataScript RT Buffer, 0.6 µL of 50X aminoallyl-dUTP/dNTP mix, 3 µL of StrataScript RNase H⁻ RT, 0.4 µL of RNase inhibitor, and 3 µL of 0.1 *M* DTT to give a final reaction volume of 30 µL.
2. Incubate at room temperature for 10 min before transferring to 42°C for 110 min.

3.4.1.3. RNA HYDROLYSIS

The RNA is removed from the cDNA reaction by alkaline hydrolysis.

1. To each reaction, add 10 µL of 1 *N* NaOH and 10 µL of 0.5 *M* EDTA (pH 8.0) and incubate at 65°C for 1 h.
2. After incubation, neutralize the reaction by adding 25 µL of 1 *M* HEPES, pH 7.5.

3.4.1.4. CLEANUP OF cDNA SYNTHESIS REACTIONS

1. Assemble the Microcon filters onto their collecting tubes. Fill each Microcon chamber with 275 µL of H_2O, add sample (approx 75 µL), and rinse the reaction

tubes with 100 µL of H$_2$O such that the total amount of H$_2$O added to each Microcon is 450 µL. Place the Microcon assemblies into a benchtop microfuge rotor such that the cap straps are aligned toward the center of the rotor, and centrifuge at 14,000g for 7 to 8 min. The aim is to reduce the volume in the upper chamber to 10–50 µL. If the volume is >50 µL, continue centrifuging for an additional 1 to 2 min and recheck the volume until it is sufficiently reduced. Should the sample spin dry, it can be recovered by adding 30 µL of H$_2$O to the membrane and agitating briefly. Note that the centrifuge times are approximate and will vary (*see* **Note 8**).
2. Perform two washes of the cDNA on the Microcon column. Once the sample volume is sufficiently reduced, discard the flowthrough and add 450 µL of H$_2$O to the upper chamber. Recentrifuge at 14,000g until the sample volume has reduced to 10–50 µL and then discard the flowthrough. Perform a second wash step.
3. Invert the Microcons and place in fresh collecting tubes. Centrifuge at 1000g for 3 min to elute. Dry the samples in a rotary evaporator. The dried cDNA samples can be stored at –20°C for several months.

3.4.1.5. Coupling Reaction

Only perform the coupling reaction when you are prepared to proceed directly to the hybridization step.

1. Resuspend the cDNA pellet in 9 µL of 0.1 M sodium bicarbonate, pH 9.0, and incubate for approx 5 min at 37°C.
2. Vortex the samples to ensure complete resuspension of the cDNA.
3. Add 1.25 µL of either Cy3 or Cy5 dye resuspended in DMSO to each cDNA sample and mix.
4. Incubate at room temperature for 1 h in the dark.

3.4.1.6. Removal of Unincorporated Dyes

The unincorporated dyes are removed from the coupling reaction using a QIA-Quick PCR Purification Kit.

1. Make each sample volume up to 100 µL by adding 90 µL of H$_2$O. Add 500 µL of buffer PB to each sample and apply to the QIA-Quick column. Centrifuge at 15,700g in a microfuge for 30–60 s and discard the flowthrough.
2. Add 750 µL of PE buffer to the column and centrifuge at 15,700g for 30–60 s. Discard the flowthrough, repeat the wash step, and discard the flowthrough again. Centrifuge the columns for an additional minute to remove any traces of PE buffer. The filters, which contain the coupled cDNA, should look pink for the Cy3 reactions and blue for the Cy5 reactions at this point.
3. Transfer the columns to fresh microfuge tubes and add 30 µL of EB buffer (10 mM Tris, pH 8.5) directly to the filter. Elute the sample by incubating for 1 min and then centrifuge at 15,700g for 1 min. Add an additional 30 µL of EB buffer to

E. coli Spotted Double-Strand DNA Microarrays

the column and repeat the elution procedure. This will give a final eluate volume of 60 µL. The samples should be pink following successful labeling for the Cy3 cDNA reactions and blue for the Cy5 cDNA reactions.

4. Pool the appropriate sample pairs (Cy3 + Cy5) to give 120 µL of purple solution. The samples are concentrated by applying them to Microcon-30 filters and centrifuging at 14,000g for ~2 min until their volume is reduced to 2–5 µL. Invert the Microcon columns in fresh collecting tubes and elute the samples by centrifuging at maximum speed for 30 s. The samples can then be dried in a rotary evaporator in the dark and stored in the dark at 4°C overnight.

3.4.2. Direct Labeling

3.4.2.1. ANNEALING

1. In microfuge tubes, mix 16–25 µg of total RNA with 10 µg of random hexamers and DEPC-treated H_2O to a final volume of 14.8 µL.
2. Incubate the mixture at 70°C for 10 min and then chill on ice for up to 10 min (*see* **Note 5**).

3.4.2.2. cDNA SYNTHESIS REACTION (*SEE* **NOTE 6**)

1. Add to each annealed RNA/random hexamer mixture 3 µL of the dNTP mix, 6 µL of 5X first-strand buffer (Gibco-BRL), 3 µL of 0.1 M DTT, 1.2 µL of 1 mM Cy-dye (3 or 5), and 2 µL of SuperScript II (Gibco-BRL).
2. Incubate at room temperature for 10 min before transferring to 42°C for 1 h, 50 min.

3.4.2.3. CLEANUP OF cDNA SYNTHESIS REACTIONS

1. Assemble the Microcon filters onto their collecting tubes. Fill each Microcon chamber with 320 µL of H_2O, add sample (approx 30 µL), and rinse the reaction tubes with 100 µL of H_2O such that the total amount of H_2O added to each Microcon is 450 µL. Place the Microcon assemblies into a benchtop microfuge rotor such that the cap straps are aligned toward the center of the rotor, and centrifuge at 14,000g for 7 to 8 min. The aim is to reduce the volume in the upper chamber to 10–50 µL. If the volume is >50 µL, continue centrifuging for an additional 1 to 2 min. Should the sample spin to dryness, it can be recovered by adding 30 µL of H_2O to the membrane and agitating briefly. Note that the centrifuge times are approximate and will vary (*see* **Note 8**).
2. Perform two washes of the cDNA on the Microcon column. Once the sample volume is sufficiently reduced, discard the flowthrough and add 450 µL of H_2O to the upper chamber. Centrifuge at 14,000g until the sample volume has reduced to 10–50 µL and then discard the flowthrough. Perform a second wash step. The samples should be concentrated to approx 5 µL after the final wash.
3. Invert the Microcon spin columns and place into fresh collecting tubes. Centrifuge at 1000g for 3 min to elute. The probes obtained by the direct incorporation protocol at this point can be used immediately for hybridization on microarrays in **Subheading 3.5.** or stored overnight in the dark at 4°C.

3.5. Sample Hybridization to Array

Two protocols are presented for microarray hybridization. One uses conventional cover slips while the second uses specialized LifterSlip brand cover slips. The conventional and LifterSlip methods are similar and are presented with the points of difference noted. The decision to employ one method or the other is a matter of preference.

3.5.1. Preparation of Sample and Array

3.5.1.1. PREPARATION OF THE SAMPLE

The hybridization mix contains from 16 to 25 µg of cDNA primed from total RNA and 20 µg of yeast tRNA (or 15 µg of poly[d(I-C)]), 2.5X SSC, 25 mM HEPES [pH 7.0], 0.225% SDS). The total volume of the hybridization should be 16 µL when using 22 × 22 mm flat cover slips and 21 µL when using 22 × 25 mm LifterSlips.

When working with aminoallyl-coupled dried probes, first thoroughly resuspend the Cy3/Cy5-labeled cDNA in 10.5 µL of H_2O if using flat cover slips or 14.1 µL if using LifterSlips and incubate at 65°C for 1 to 2 min to aid resuspension. If using direct-labeled probes, mix the Cy3- and Cy5-labeled cDNAs at room temperature and adjust the volume to 10.5 or 14.1 µL for flat or LifterSlips, respectively.

1. Flat cover slip protocol:
   a. To the 10.5-µL probe mixture (it is important to ensure that the Cy3 and Cy5 probes are completely and thoroughly mixed), add 2 µL of 20X SSC, 2 µL of 10 mg/mL tRNA, and 0.8 µL of 500 mM Hepes (pH 7.0).
   b. Mix by flicking after adding each component.
   c. Add 0.75 µL of 5% SDS immediately before denaturing the probe in **step 2b**.
2. LifterSlip protocols:
   a. To the 14.1 µL of probe mixture add 3.3 µL of 20X SSC, 2.5 µL of poly(dI-dC), and 1.1 µL of 0.5 M HEPES (pH 7.0) to give a final volume of 21 µL. SDS is added to the LifterSlip hybridization mix after filtration in the following step.
   b. To remove any particulates, the samples are filtered using UltraFree-MC Low Binding 0.45-µm filter columns. Prewet the filter membranes by adding 10 µL of H_2O and centrifuge at 7000g for 2 min in a microcentrifuge. Remove the flowthrough and deposit the hybridization mix as two droplets on the inside of the filter case. Centrifuge the samples at 15,000g for 2 min, and collect the filtrate in 0.5-mL microfuge tubes containing 1 µL of 5% SDS. Take care to accurately measure the SDS; excess SDS in the samples may precipitate.

3.5.1.2. DENATURING OF SAMPLES

1. Incubate the samples at 95–100°C for 2 min.
2. Allow the samples to cool briefly after incubation and centrifuge to collect.

E. coli Spotted Double-Strand DNA Microarrays

3.5.1.3. PREPARATION OF ARRAY
1. While the samples are cooling, place the array slides in the hybridization chambers, taking care to remove any dust using a stream of compressed air.
2. If using LifterSlips clean them carefully with high-grade ethanol, and then dry and dust them with compressed air; this step aids the application of the hybridization sample to the array and reduces the chance of bubbles forming under the LifterSlip.
3. Carefully place a LifterSlip over each microarray such that the dull white strips are along the long axis of the slide and touching the glass.
4. To maintain humidity inside the hybridization chamber during the incubation, apply a 3-µL drop of 3X SSC to each corner of the slide.

3.5.2. Sample Application and Hybridization
1. LifterSlip protocol:
    a. Apply the hybridization mix to the array by placing a pipet tip at one end of the LifterSlip and allow the sample to move up underneath the cover slip by capillary action.
    b. Move the pipet tip repeatedly along the length of the cover slip to avoid any bubbles.
    c. Once the space beneath the LifterSlip is completely full, add sample to the other end of the LifterSlip to top off both ends.
2. Flat cover slip protocol:
    a. Apply the hybridization mix to the array by pipeting it gently onto the center of the array taking care not to introduce air bubbles.
    b. Using tweezers apply a cover slip on top of the hybridization mix. Place the edge of the cover slip down first just beyond the edge of the array and then allow it to fall from 45° or less to the slide.
3. Place the cover on the hybridization chamber and tighten the lid screws to make watertight. Place the chamber into a water bath and incubate at 63–65°C for at least 5 to 6 h or overnight (12 h maximum). Take care to keep the chamber horizontal at all times so as not to disturb the 3X SSC droplets or the cover slip.

3.5.3. Stringency Wash Step
1. After incubation, remove the chambers one at a time from the water bath and dry each with a paper towel. Keep the chambers level at all times and handle them gently in order not to disturb the cover slips. Unscrew the chamber and remove the slide.
2. For the first rinse, wash the slides one at a time by placing each on a slide rack immersed in a glass dish containing 330 mL of distilled H_2O, 20 mL of 20X SSC, and 1 mL of 10% SDS. Move the slide gently up and down in order to dislodge the coverslip, taking care not to allow it to scratch the surface of the array. Plunge the rack up and down in the solution 10–20 times.
3. For the second rinse, to avoid carryover of SDS from the first wash, blot the base of each slide on paper towels before transferring to a slide rack in a second glass

dish containing 350 mL of distilled H$_2$O and 1 mL of 20X SSC. However, do not allow the surface of the slide to dry completely during the transfer between washes. After all slides are in the second rack, plunge the rack a few times in the wash solution and then remove the entire rack and blot on a paper towel. Dry the arrays by centrifuging at 100g for 5 min (*see* **Note 9**).
4. Scan the arrays within several hours of hybridization—the fluorophor signal deteriorates with time.

3.6. Normalization Procedure

Following image acquisition and spot assignment using a commercially or freely available software package (*see* **Note 10**), the signal intensities in both channels should be normalized according to an *a priori* experimental design. Normalization is a key step in data processing that can significantly affect later analyses. If improperly carried out, it can lead to significant artifacts in gene expression patterns.

3.6.1. Average Bulk Normalization

One of the more common methods of normalization is to equalize the cumulative intensities of the Cy3- and Cy5-labeled channels. This method is based on the assumptions that (1) total RNA levels are constant between the two samples; and that (2) expression changes in the experimental channel will consist of equal, or almost equal, proportions of over- and underexpressed mRNAs.

To accomplish average bulk normalization, first select a representative set of spots from the microarray based on a combination of criteria including signal intensity, homogeneity of spot color ratio, and fraction of pixels above background. Ensure that the selected set is representative of the entire array; that is, at least 50% of all spots should be included in the selected set.

The normalization factor is calculated by taking the logarithm of the red-to-green intensity ratios observed for each spot in the selected representative set. The sum of these logarithms, according to the assumptions of average bulk normalization, should be equal to 0. Thus, subtracting from each logged ratio the average of the logs of ratios will satisfy the requirement for the sum. This average of the logarithms of ratios is the normalization factor.

3.7. General Considerations for Controlling Microarray Quality

Highly parallel analysis of nucleic acid abundances on DNA microarrays presents new challenges in experimental quality control. We find that many elements of the experiment must be tuned simultaneously to obtain reliable data from these studies, including, but not limited to, hybridization conditions, DNA spotting, and normalization. Failure to adjust correctly any combination of experimental parameters can lead to the production of misleading results.

Following are several points to consider in assessing the quality of data obtained in two-color microarray hybridizations:

1. Any spots from which data are obtained must be representative for that array, and abnormal or non-uniform spots should be visually flagged and removed from further analysis. If a large fraction of spots on the array must be flagged, the quality of the remaining data is suspect.
2. If the experimental design calls for average bulk normalization (*see* **Subheading 3.6.1.**), verify that the normalized ratios are normally distributed. While estimating a fit to a Gaussian distribution may not always be a practical option, one should be alert for the warning signs such as bi- or multimodality, disparity between the mean and median ratios, and more than a twofold excess of ratio values falling into the distribution tails. Finally, if normalization was performed on a properly representative set of genes, the distribution of the ratios in the log space for the entire array should be centered at or very near zero. If you observe any of these deviations from expectations, repeat the normalization by including more spots in the normalization set. If the problems do not resolve, exclude the array from further analysis.
3. To ensure the integrity of microarray data, one should also control for nonspecific hybridization to the array. This is accomplished by spotting elements on the array that show no appreciable sequence identity to any of the potential probes. We have used 384 yeast amplicons for this purpose. Although some hybridization signal is generated by these elements, it has been consistently lower than the signal from the ORF elements on array. Ideally, only spots that demonstrate signal intensity reproducibly above the signal intensity of the yeast control spots should be included in the analysis of *E. coli* data.
4. It is important to have a consistent and relatively large amount of DNA spotted on each element in the array. The amount of DNA in a spot available for hybridization is influenced by the concentration of DNA in the printing solution, the salt concentration in the printing solution, the printing mechanism, and the quality of the slide surface. We have observed that printing from 3X SSC DNA solutions containing 100–300 ng/μL of nucleic acid produces acceptable spot quality. Spots printed in this manner provide maximal fluorescence intensity after 6 h of hybridization and demonstrate independence between expression ratios and variation in the amount of deposited DNA.

3.8. E. coli Internet Resources

A large class of bacterial microarray experiments involves searching for genes whose RNA levels change on some perturbation to the culture, whether it is a different phase of growth, a genetic change, or a change to the medium or environment. First, one determines the set of genes that is perturbed under a given condition. Second, one wishes to solve the puzzle of how the condition causes those perturbations. In facing this challenge, it is essential to have access to comprehensive information about the genes of an organism and their potential relationships. In the following sections, we describe several Internet-accessible *E. coli*–oriented databases that are useful in the analysis and interpretation of microarray data.

3.8.1. NCBI

There are three genome sequences for *E. coli* at NCBI. These are *E. coli* K12 (October 13, 1998) *(9)*, *E. coli* O157:H7 (March 7, 2001), and *E. coli* O157:H7 EDL933 (January 25, 2001). Links to the complete genome sequences for *E. coli* and other bacteria can be found at http://www.ncbi.nlm.nih.gov/PMGifs/Genomes/eub.html. At the NCBI site, one can search for genes by name or position in the genome. Once a gene is identified, it can be displayed in a genome browser in which other nearby genes are displayed, color coded by metabolic function. By clicking on a gene in the genome browser, one arrives at a page where the best BLAST hits to the gene are identified and displayed. From this page, links to other information about the gene and its BLAST neighbors are available, including taxonomy, conserved domains, and clusters of orthologous groups. The NCBI site is well linked and its sequence data are updated frequently. Work on improving the site is ongoing, and the appearance of new views and tools is frequent.

3.8.2. Colibri

The Colibri Web site at http://genolist.pasteur.fr/Colibri, like the NCBI site, provides a graphic genome browser. This site provides the capability to identify genes by name or to search within chromosomal regions. Features at this site include the ability to search for genes by their codon usage, molecular weight, and isoelectric point. One can easily view and download sequences (protein and DNA) and gene lists for specified regions of the genome. Lists of genes can be created by searches and then sorted by various criteria. The genome represented is that of *E. coli* K12 MG1655 *(10)*, and the Colibri annotation corresponds to that of EcoGene.

3.8.3. EcoGene

The EcoGene Web site contains a wealth of useful information and annotation for *E. coli.* (http://bmb.med.miami.edu/EcoGene/EcoWeb) *(11)*. Information at the site is developed in collaboration among K. Rudd, Colibri, and the *E. coli* Genetic Stock Center at Yale (http://cgsc.biology.yale.edu) *(9)*. The sequence data at the site are that of *E. coli* K-12. Among the most important links at this site are those to the most relevant literature for each gene. The ability to identify articles about genes through curated selections can be more direct than using keyword-based search strategies.

3.8.4. GenProtEC

The GenProtEC Web site (http://geneprotec.mbl.edu) *(12)* provides information about functional gene categories in *E. coli*. There are many ways in which the site can be queried; one can go from gene to category and category to gene. This site is a good place to look for functional relationships among genes. The categories are described in a three-level hierarchy, which makes learning about either the broad cellular role or the specific function of a protein straightforward.

3.8.5. RegulonDB

RegulonDB (http://www.cifn.unam.mx/Computational_Genomics/regulondb) *(13)* is a database of transcriptional regulation in *E. coli*. A large number of regulators and their regulatory loci can be found at its site. RegulonDB is also part of the EcoCyc database, and it provides a genome browser in which operons are marked.

3.8.6. EcoCyc

EcoCyc (http://www.ecocyc.org) *(14)* is a comprehensive Web site for *E. coli* gene and pathway information. The site contains a database of genes, enzymatic reactions and pathways, functional categories *(12)*, and transcriptional regulators (RegulonDB) *(13)*. Notably, at EcoCyc it is possible to map expression values onto pathways. One drawback of the site is that it is difficult to download/export large amounts of information, something often required in array-based investigations. EcoCyc is updated frequently and new tools come online periodically.

3.8.7. E. coli *Entry Point*

The *E. coli* Entry Point (http://coli.berkeley.edu/ecoli) is an effort to organize and provide access to information about *E. coli* in a way that makes it easy to interpret and understand microarray data. The site facilitates analysis of microarray results by allowing the identification of relationships among lists of genes. Among the relationships that can be identified are proximity on genome, common operon, sequence similarity, common transcriptional regulator, and common gene function. Genes can be sorted by location on the genome and grouped by operon. One can identify known nearby regulatory protein-binding sites common to a list of genes and display these in a table. Hyperlinks to external data sources are provided for access to more detailed information about individual genes. All of the information at the *E. coli* Entry Point was downloaded from external sources.

4. Notes

1. To avoid ribonuclease contamination during RNA isolation and fluorescent labeling, use only RNase-free and/or DEPC-treated solutions, RNase-free microfuge tubes, and filter pipet tips. For storing and handling of RNase-free solutions, we recommend using sterile, disposable polypropylene tubes. These tubes are generally RNase free and do not require further treatment. Glassware, however, should be DEPC treated and baked to ensure that it is RNasefree.
2. At this point in the protocol cells can alternatively be resuspended and RNA extracted according to the RNeasy Midi protocol by Qiagen. Be sure to include the "optional" on-column DNase treatment step described in the manual. RNA can be isolated from bacterial cells by either chemical extraction or binding to a silica gel–based membrane. Factors to consider in choosing a method include cost of materials, time constraints, toxicity of reagents, and personal preference. Advantages of the chemical extraction protocol presented include scalability for greater or smaller yield and lower cost. Advantages of using RNA binding membranes such as RNeasy from Qiagen are speed, lower chance of ribonuclease contamination, and lower toxicity of reagents. Experience has shown that RNA preparation is more consistent using the RNA binding column method provided that kits are less than 6 mo old. The column method has the additional advantage of allowing on-column DNase digestion. This avoids the second purification step necessary to inactivate and remove the nuclease if digestion is performed separately.
3. This protocol allows for a quick-chill quenching of bacterial metabolism and does not involve reagents that potentially perturb the polyadenylation reaction and/or the integrity of ribosomes and subribosomal particles.
4. The mechanical lysis procedure is efficient only using a Mini-BeadBeater and has not been reproduced adequately when vigorous vortexing was used instead.
5. Only the direct-labeling protocol has been attempted with enriched mRNA. All RNA obtained at the end of the mRNA isolation and enrichment protocol should be used in a single labeling reaction.
6. We estimated the cost of one reaction (without bulk discounts and promotion prices) using the direct incorporation protocol to be $57.00; the cost of indirect labeling is estimated at $31.00 per reaction.
7. This method has been adopted from the Web site www.microarrays.org, maintained by Dr. Joe DeRisi at the University of California at San Francisco.
8. About 1 of every 25 Microcon YM-30 filters depending on the batch does not retain the labeled cDNA. An indication of filter failure when following the direct-labeling protocol is rich color of the flowthrough and/or dry membranes on individual filters of a set, while most of the other filter membranes remain wet. In the event of such a failure, the sample can be recovered by applying the flowthrough to a new filter.
9. To avoid streaking on the slide surface, this step has to be performed swiftly. Use only centrifuges that provide rapid acceleration (Beckman models GP and Allegra 6, or comparable).

10. We currently use series 4000 scanners from Axon for image acquisition and the accompanying GenePix software for image analysis. We have also used the free ScanAlyze software package by Mike Eisen for image analysis, which is available for download at http://rana.lbl.gov.

Acknowledgments

We wish to express our gratitude for intellectual, emotional, and financial support to Pat Brown, David Botstein, Stanley Cohen, Nick Cozzarelli, Carol Gross, and Sydney Kustu. Work in the laboratory of Pat Brown was supported by the Howard Hughes Medical Institute and National Institutes of Health (NIH) grant HG00938; in the laboratory of Nick Cozzarelli by NIH grant GM31657; in the laboratory of Sydney Kustu by NIH grant GM38361 and from Novartis Insitute; in the laboratory of Carol Gross by NIH grant GM30477. JAB was supported by NIH grant GM 07790, VFW by Deutsche Forschungsgemeinschaft, and DPZ by NIH Fellowship GM19862.

References

1. Courcelle, J., Khodursky, A., Peter, B., Brown, P. O., and Hanawalt, P. C. (2001) Comparative gene expression profiles following UV exposure in wild-type and SOS-deficient Escherichia coli. *Genetics* **158,** 41–64.
2. Khodursky, A. B., Peter, B. J., Schmid, M. B., DeRisi, J., Botstein, D., Brown, P. O., and Cozzarelli, N. R. (2000) Analysis of topoisomerase function in bacterial replication fork movement: use of DNA microarrays. *Proc. Natl. Acad. Sci. USA* **97,** 9419–9424.
3. Wendisch, V. F., Zimmer, D. P., Khodursky, A., Peter, B., Cozzarelli, N., and Kustu, S. (2001) Isolation of Escherichia coli mRNA and comparison of expression using mRNA and total RNA on DNA microarrays. *Anal. Biochem.* **290,** 205–213.
4. Zimmer, D. P., Soupene, E., Lee, H. L., Wendisch, V. F., Khodursky, A. B., Peter, B. J., Bender, R. A., and Kustu, S. (2000) Nitrogen regulatory protein C-controlled genes of Escherichia coli: scavenging as a defense against nitrogen limitation. *Proc. Natl. Acad. Sci. USA* **97,** 14,674–14,679.
5. Lindahl, T. and Andersson, A. (1972) Rate of chain breakage at apurinic sites in double-stranded deoxyribonucleic acid. *Biochemistry* **11,** 3618–3623.
6. Lindahl, T. and Nyberg, B. (1972) Rate of depurination of native deoxyribonucleic acid. *Biochemistry* **11,** 3610–3618.
7. Feng, Y. and Cohen, S. N. (2000) Unpaired terminal nucleotides and 5′ monophosphorylation govern 3′ polyadenylation by Escherichia coli poly(A) polymerase I. *Proc. Natl. Acad. Sci. USA* **97,** 6415–6420.
8. Sambrook, J., Maniatis, T., and Fritsch, E. F. (1989) *Molecular Cloning: A Laboratory Manual*, 2nd ed. Cold Spring Harbor Laboratory, Cold Spring Harbor, NY.
9. Berlyn, M. B. and Letovsky, S. (1992) Genome-related datasets within the E. coli Genetic Stock Center database. *Nucleic Acids Res.* **20,** 6143–6151.

10. Blattner, F. R., Plunkett, G. 3rd, Bloch, C. A., et al. (1997) The complete genome sequence of Escherichia coli K-12. *Science* **277,** 1453–1474.
11. Rudd, K. E. (2000) EcoGene: a genome sequence database for Escherichia coli K-12. *Nucleic Acids Res.* **28,** 60–64.
12. Riley, M. and Labedan, B. (1996) *E. coli* gene products: physiological functions and common ancestries, in *Escherichia coli and Salmonella: Cellular and Molecular Biology*, 2nd ed., vol. 2. (Neidhardt, F., Curtiss, R. III, Lin, E. C. C., Ingraham, J., Low, K. B., Magasanik, B., Reznikoff, W., Riley, M., Schaechter, M., and Umbarger, H. E., eds.). ASM Press, Washington, DC, pp. 2118–2202.
13. Salgado, H., Santos-Zavaleta, A., Gama-Castro, S., Millan-Zarate, D., Diaz-Peredo, E., Sanchez-Solano, F., Perez-Rueda, E., Bonavides-Martinez, C., and Collado-Vides, J. (2001) RegulonDB (version 3.2): transcriptional regulation and operon organization in Escherichia coli K-12. *Nucleic Acids Res.* **29,** 72–74.
14. Karp, P. D., Riley, M., Saier, M., Paulsen, I. T., Paley, S. M., and Pellegrini-Toole, A. (2002) The EcoCyc DB. *Nucleic Acids Res.* **30(1),** 56.

6

Isolation of Polysomal RNA for Microarray Analysis

Yoav Arava

1. Introduction

DNA microarrays have been used extensively in recent years to study mRNA expression profiles of different cell types under various growth conditions. These steady-state mRNA profiles provide a wealth of information about cellular functions and responses. However, they do not necessarily reflect the ultimate gene expression profile of a cell since the step of translation might lead to discrepancies between the mRNA profile and the profile of the actual functioning unit in the cell, the protein *(e.g., 1,2)*.

One of the most common methods to study translation mechanisms and regulation is polysomal profiling using sucrose gradients. The method involves size separation of large cellular components on a sucrose gradient and monitoring the A_{260} across the gradient. Although termed *polysomal profile*, the A_{260} profile of the separated complexes (**Fig. 1**) contains information not only about polyribosomes (two or more ribosomes), but also about other translation machinery components, including the small ribosomal subunit, large ribosomal subunit, and single ribosome. Changes in this profile are indicative of changes in translation. For example, yeast cells growing under different carbon sources present different polysomal profiles owing to inhibition of translation initiation *(3)*.

To study the behavior of specific mRNAs, the gradient is fractionated and the different fractions are analyzed by Northern analysis. Important information regarding translation of these mRNAs is obtained. The fraction of translationally active messages (those associated with ribosomes) can be deduced from the ratio between messages that sediment in the polysomal fractions vs those that sediment in the nonpolysomal fractions. Changes in this ratio indicate

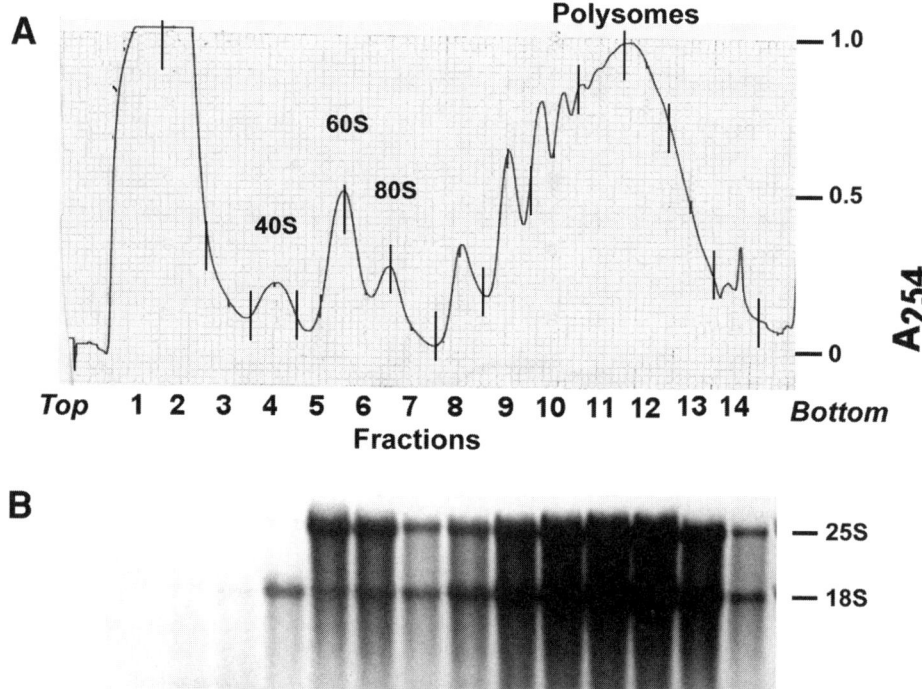

Fig. 1. Representative polysomal profile obtained from yeast cells grown to mid-log phase in rich medium. (**A**) Continuous OD_{254} of 10–50% sucrose gradient with peaks corresponding to the 40S, 60S, 80S, and polysomal complexes indicated. (**B**) The gradient was separated to 14 fractions and run on an agarose gel. Fraction 4 contains the 40S subunit, as apparent by the strong 18S rRNA band, whereas fraction 5 (60S) contains mostly the 25S rRNA band. Fractions 7 and higher contains equal amount of both 18S and 25S, indicating intact ribosomes. Note the variable amounts of mRNA and rRNA between fractions.

changes in translation efficiency, as was demonstrated for the L32 ribosomal mRNA *(4)*. Another important indication of translation regulation comes from analyzing changes in the number of ribosomes with which the mRNA is associated. For example, the yeast transcription regulator GCN4 normally sediments in the fraction containing only one ribosome, but upon amino acids starvation, it appears to be associated with more ribosomes, indicative of translation activation *(5)*.

The use of DNA microarrays enables simultaneous analysis of thousands of genes from each polysomal fraction. Such analyses were performed for yeast cells *(6)* and for mammalian cells *(7–9)* and led to unsupervised identification of many translationally regulated genes.

Isolation of Polysomal RNA

The following protocol is designed to obtain sufficient amounts of high-quality mRNA for microarray analysis. By using this protocol, a clear and reproducible signal for even very low abundance yeast genes is observed.

2. Materials (see Note 1)

1. Cycloheximide: Dissolve in water to 10 mg/mL, and store at –20°C up to several months. Protect from light. Cycloheximide is highly toxic, with an LD_{50} of <50 mg/kg (see **Note 2**).
2. Heparin: Dissolve in water to 10 mg/mL and store at –20°C (see **Note 3**).
3. 75% Sucrose (w/v): This is the stock solution for preparation of the gradients. Dissolve in water and filter.
4. 10% Sucrose: 10% sucrose, 20 mM Tris-HCl, pH 8.0, 140 mM KCl, 5 mM MgCl$_2$, 0.5 mM dithiothreitol (DTT), 0.1 mg/mL of cycloheximide, and 0.5 mg/mL of heparin. Vortex well. Prepare fresh each time, 3 mL for every gradient.
5. 20% Sucrose: 20% sucrose, 20 mM Tris-HCl, pH 8.0, 140 mM KCl, 5 mM MgCl$_2$, 0.5 mM DTT, 0.1 mg/mL of cycloheximide, and 0.5 mg/mL of heparin. Vortex well. Prepare fresh each time, 3 mL for every gradient.
6. 30% Sucrose: 30% sucrose, 20 mM Tris-HCl, pH 8.0, 140 mM KCl, 5 mM MgCl$_2$, 0.5 mM DTT, 0.1 mg/mL of cycloheximide, and 0.5 mg/mL of heparin. Vortex well. Prepare fresh each time, 3 mL for every gradient.
7. 40% Sucrose: 40% sucrose, 20 mM Tris-HCl, pH 8.0, 140 mM KCl, 5 mM MgCl$_2$, 0.5 mM DTT, 0.1 mg/mL of cycloheximide, and 0.5 mg/mL of heparin. Vortex well. Prepare fresh each time, 3 mL for every gradient.
8. 50% Sucrose: 50% sucrose, 20 mM Tris-HCl, pH 8.0, 140 mM KCl, 5 mM MgCl$_2$, 0.5 mM DTT, 0.1 mg/mL of cycloheximide, and 0.5 mg/mL of heparin. Vortex well. Prepare fresh each time, 3 mL for every gradient.
9. Yeast lysis buffer: 20 mM Tris-HCl, pH 8.0, 140 mM KCl, 1.5 mM MgCl$_2$, 0.5 mM DTT, 0.1 mg/mL of cycloheximide, 1 mg/mL of heparin, and 1% Triton X-100. Prepare fresh each time and cool on ice.
10. Mammalian lysis buffer: 15 mM Tris-HCl, 5 mM MgCl$_2$, 0.3 M NaCl, 0.5 mM DTT, 0.1 mg/mL of cycloheximide, 1 mg/mL of heparin, and 1% Triton X-100. Prepare fresh each time and cool on ice.
11. Phosphate-buffered saline (PBS) containing 0.1 mg/mL of cycloheximide.
12. Pump solution: 40% sucrose, 0.01% bromophenol blue in water. This solution is used to fill the syringe pump and to push the gradient through the UV detector. The addition of bromophenol blue enables clear detection of boundary between the sample and the pump solution.
13. 8 M Guanidine HCl.
14. 7.5 M LiCl: Prepare with diethylpyrocarbonate (DEPC)-treated water, filter, and keep at room temperature.
15. TE: 10 mM Tris-HCl, pH 8.0, 1 mM EDTA, pH 8.0.
16. Water-saturated phenol:chloroform: Mix equal volumes of water-saturated phenol with chloroform. Keep at 4°C protected from light.
17. Beckman polyallomer centrifuge tubes (14 × 89 mm; cat. no. 331372).

18. Glass beads (0.45–0.55 mm diameter). Cool to 4°C before use.
19. Collection system: The gradients are separated and collected using a collection system that includes the following components (a complete description of the components and manuals can be obtained from the suppliers: Isco at www.isco.com, and Brandel at www.brandel.com):
    a. ISCO UA-6® absorbance detector with a 254-nm type 11 filter.
    b. ISCO Retriever II fraction collector.
    c. Brandel syringe pump (SYR-101).
    d. Brandel model no. 184 fractionator with 5-mm flow cell.
20. BOREX™ glass tubes (13 × 100 mm).

3. Methods

3.1. Preparation of Gradient

Linear sucrose gradients are prepared by layering different concentrations of sucrose and storing the gradients overnight at 4°C. During this incubation, the sucrose diffuses and generates a linear gradient. Alternatively, a gradient maker can be used to prepare the gradients (in this case only a short incubation at 4°C is necessary). The volumes indicated are for one gradient of 11 mL and should be increased proportionally for every additional gradient.

1. Prepare 3-mL mixes of 10, 20, 30, 40, and 50% sucrose solutions in 15-mL tubes. Vortex well.
2. Using a long Pasteur pipet, layer 2.2 mL (2.2 mL is up to the neck of the pipet) of each 10, 20, 30, 40, and 50% sucrose solution in Beckman polyallomer tubes. I find that the easiest way is to underlay the solutions: first layer the 10% sucrose and then underlay the 20% sucrose by inserting the tip of the pipet almost to the bottom of the tube and slowly inject the 20% solution under the 10% solution. Try to avoid mixing of the two layers or introducing air bubbles (*see* **Note 4**). Underlay the 30, 40, and 50% sucrose solutions in the same manner. Keep the remnants of the solutions to use for setting the baseline (*see* **Subheading 3.3., step 1**).
3. Cover with aluminum foil and store overnight at 4°C to allow equilibration.

3.2. Cell Lysis

3.2.1. Lysis of Yeast Cells

The following protocol uses mild detergent and vortexing in the presence of glass beads to lyse yeast cells. The volumes are for an 80 mL culture (one gradient). Extracts are routinely made from 500 mL of culture and split into six gradients. It is important to perform all steps on ice with precooled solutions and tubes.

1. Start an overnight culture and grow cells to mid–log phase (OD_{600}: 0.4–0.6). Doubling time for strains S288c and BY4741 growing in YPD at 30°C is

approx. 90 min. The doubling time will vary depending on the strain and growth conditions.
2. Add cycloheximide to a final concentration of 0.1 mg/mL, cool, and immediately spin down the cells at 6000g for 4 min at 4°C (*see* **Note 5**).
3. Discard the supernatant, resuspend the cell pellet in 2.5 mL of lysis buffer, and spin as in **step 2**.
4. Repeat **step 3**.
5. Resuspend in 0.7 mL of lysis buffer, transfer to corex tubes, and add approx 0.6 vol of chilled glass beads (0.45–0.55 mm). Vortex hard for 20 s and cool on ice for 100 s. Repeat four times to achieve complete lysis.
6. Spin down at 2600g for 5 min at 4°C. Transfer the supernatant (approx. 500 µL) to a 1.6-mL tube.
7. Spin at 7200g for 5 min at 4°C. Transfer to a new 1.6-mL tube.
8. Bring to a final volume of 1 mL with lysis buffer. Carefully load ~0.8 mL of the lysate on each gradient.
9. Insert the gradients into precooled SW41 rotor buckets. If necessary, balance the gradients with lysis buffer (*see* **Note 6**).
10. Spin at 35,000 rpm for 160 min at 4°C (*see* **Note 7**).

3.2.2. Lysis of Mammalian Cells

This protocol was used for HeLa cells grown in suspension. For adherent cells, use the same solutions and lyse the cells directly on the plate.

1. For each gradient, grow 2×10^7 cells (at a density of 2×10^5 cells/mL).
2. Add cycloheximide to a final concentration of 0.1 mg/mL. Immediately spin at 200g for 4 min.
3. Wash the cells twice with 5 mL of cold PBS containing 0.1 mg/mL of cycloheximide.
4. Resuspend in 1 mL of mammalian lysis buffer.
5. Incubate on ice for 10 min with occasional mixing.
6. Spin at 12,000g for 10 min. Transfer to a new tube.
7. Carefully load 0.8 mL of the lysate on each gradient.
8. Insert the gradients into precooled SW41 rotor buckets. If necessary, balance the gradients with lysis buffer (*see* **Note 6**).
9. Spin at 35,000 rpm for 190 min at 4°C (*see* **Note 7**).

3.3. Collection of Fractions

There are multiple methods to collect fractions from the sucrose gradient. The simplest one involves puncturing the bottom of the tube and collecting drops. OD_{260} measurement of each fraction will give an estimate of the amount of nucleic acid in each fraction. The method described here uses a syringe pump to push the gradient through a UV detector, thereby generating a continuous OD profile in which the first fractions contain the slow migrating complexes (**Fig. 1**).

Gradients are fractionated and collected using the ISCO collection system with the following settings: pump speed, 0.75 mL/min; fraction time, 1.2 min/fraction; chart speed, 60 cm/h; sensitivity, 1 to 2 (depending on the desired resolution); peak separator, off; noise filter, 1.5. These settings result in 14 fractions each with a volume of ~0.9 mL. They can be easily modified to collect more or fewer fractions (e.g., a fraction time of 1.5 min will lead to collection of 10 fractions, approx 1.2 mL each).

1. For baseline setting, fill the syringe with pump solution. Assemble the blank polyallomer tube containing 10% sucrose (use the remnants from **Subheading 3.1., step 2**) in the flow cell and pump it through the UV detector. Once the 10% sucrose solution gets to the UV detector, set the sensitivity dial to "Set lamp and optics" and press the "Auto baseline" button. Then switch the sensitivity dial to 2 and again press the "Auto baseline" button.
2. Wash the flow cell by reversing the pump and drawing approx 2 mL of water, then 2 mL of 70% ethanol and again 2 mL of water.
3. Remove the polysome gradients from the ultracentrifuge. Assemble one gradient and pump the sample through the detector. Keep the rest of the gradients at 4°C (*see* **Note 8**).
4. Start the retriever as the sample begins dripping. Do not collect the first one or two drops; they contain some material from the tubing (*see* **Note 9**).
5. Collect the fractions in 13 × 100 mm glass tubes containing 2 mL of 8 M guanidine HCl.
6. Collect the whole polysomal gradient, including a few drops of the pump solution (which can be easily distinguished because it is blue).
7. Wash the flow cell as in **step 2**.
8. If collecting more than one gradient, repeat **steps 3–7** for each gradient. Since similar fractions from several gradients will be pooled together in the next steps, it is important to ensure that in all gradients the same fractions contain the same polysomal complexes (*see* **Note 10**).

3.4. RNA Extraction

To obtain sufficient amounts of mRNA for microarray analysis, identical fractions from at least two gradients are pooled together. Indicated volumes are for material from three gradients.

1. Pool similar fractions into a 50-mL Oak Ridge tube, and precipitate by adding an equal volume of 100% ethanol and incubating overnight at –20°C.
2. Spin at 12,000g for 20 min at 4°C using an SS-34 rotor. Discard the supernatant.
3. Add 1 mL of 85% ethanol and spin as in **step 2**.
4. Resuspend in 400 µL of TE, pH 8.0. The first fractions contain high amounts of heparin. To ease their resuspension, add the TE and let the sample stand for a few minutes at room temperature.

Isolation of Polysomal RNA

5. Transfer to a 1.6-mL tube, and precipitate again by adding 0.1 vol of 3 M sodium acetate, pH 5.3, and 2.5 vol of 100% ethanol.
6. Resuspend in 650 µL of double-distilled water. To remove any residual proteins add 650 µL of water-saturated phenol:chloroform. Vortex vigorously and centrifuge for 2 min at top speed. Transfer 500 µL of the aqueous phase to a new tube.
7. Bring to 1 mL with water and add LiCl to a final concentration of 1.5 M. Incubate overnight at −20°C. The LiCl precipitation will remove any residual heparin, which might interfere with the labeling reaction (*see* **Subheading 3.5.**).
8. Thaw the samples at 4°C. Spin down at top speed in a cooled centrifuge for 30 min. Carefully discard the supernatant. Add 200 µL of 75% ethanol and spin again as just described. Discard the supernatant, air-dry, and resuspend in 150 µL of water.
9. To remove any residual LiCl, precipitate again by adding 0.1 vol of 3 M sodium acetate, pH 5.3, and 3 vol of 100% ethanol and incubating overnight at −20°C.
10. Spin down, wash with 75% ethanol, and air-dry as in **step 8**.
11. Resuspend in 25 µL of 1 mM Tris, pH 7.4, and store at −70°C.
12. Analyze 5 µL of each sample on 1.5% agarose gel. The profile should be similar to the one in **Fig. 1B**: a strong band corresponding to the 18S RNA at fraction 4, 25S rRNA at fraction 5, and both bands corresponding to both rRNA in fractions 6–14 (*see* **Fig. 1B**).
13. Use 14 µL of each sample for labeling.

3.5. RNA Labeling and Hybridization

RNA labeling and hybridization steps are performed according to the standard protocols (http://cmgm.stanford.edu/pbrown/protocols/index.html). Several important points should be considered regarding the polysomal profiles:

1. Different fractions contain different amounts of RNA (both total RNA and mRNA). To maintain these relative differences, equal amounts of exogenous mRNA can be spiked to each of the fractions at the first step of collection. PCR fragments complementry to these mRNA should be spotted on the microarray, and their signal can be used to normalize for differences between fractions.
2. Do not attempt to label and hybridize a large number (>5) of samples at the same time. Usually I split the labeling and hybridization of 14 microarrays into three groups.
3. If hybridizing the samples with a common, unrelated reference, it is important to use exactly the same reference sample in all fractions in order to reduce variation. In such "type 2" experiments, 15 µg of yeast RNA prepared by the hot phenol procedure is sufficient for each hybridization.
4. Labeling should be made with oligo-dT primers. The use of random primers is not recommended since most of the generated signal will be from rRNA. This nonrelevant signal might interfere with the hybridization.

4. Notes

1. There is no need to make reagents as RNase free up to **Subheading 3.4., step 6**, since all solutions contain substantial amounts of heparin. For **steps 7–13**, several precautions should be taken: Reagents should be of the highest grade possible and DEPC-treated water should be used. All plasticware (e.g., Eppendorf tubes and tips) should be RNase free. Always use gloves when using these solutions.
2. Cycloheximide inhibits the ribosome's translocation step in eukaryotic cells. Its use is crucial for obtaining sufficient amounts of polysomes, and its depletion will lead to ribosomes falling off the mRNA.
3. Heparin is a very potent enzyme inhibitor that is used here to inhibit RNases. It is added both to the lysis buffer and to the gradient and must be removed in later steps (*see* **Subheading 3.4., step 7**) to permit efficient labeling of the RNA. The use of other RNase inhibitors is more expensive and less efficient; I tried RNasin® in concentrations of up to 150 U/mL and did not get sufficient inhibition.
4. Alternatively, one can overlay the solutions by starting with the highest concentration and layering the lower concentrations by carefully sliding the solutions along the wall of the tube.
5. I have noticed that extended incubation on ice prior to centrifugation leads to an increase in the 80S fraction, most probably because free mRNAs are being loaded with ribosomes.
6. Although most rotors can tolerate up to 0.1 g differences between paired samples, I try not to exceed 0.02 g differences. Imbalances will normally cause the rotor to stop but sometimes might damage the centrifuge.
7. Detailed theoretical and practical aspects of sedimentation in sucrose gradients are described in **ref. 10**.
8. No change in the profile was observed even when the gradients were left at room temperature for several hours, indicating that the sedimented complexes are very stable and do not diffuse in the gradients.
9. There are few seconds of void volume between the OD reading and the onset of dripping into the collection tube. Therefore, the fraction mark on the OD_{254} profile made by the retriever does not match the actual collected fraction. To correct for this, each mark should be shifted to the left according to the size of void volume.
10. The A_{254} profiles of the gradients should be identical. Small variation might appear at the top of the gradient (fractions 1 and 2) owing to differences in the volume of the loaded material. It is important to verify that the same fractions contain the same ribosomal complexes, because they will be pooled together in later steps.

Acknowledgments

Thanks to Ken Kuhn, Yulei Wang, Matt L. Peck, and Julie B. Sneddon for technical advise and protocols. Y.A. is a research associate in the Howard Hughes Medical Institute.

References

1. Clemens, M. J. and Bommer, U. A. (1999) Translational control: the cancer connection. *Int. J. Biochem. Cell. Biol.* **31(1)**, 1–23.
2. Krichevsky, A. M., Metzer, E., and Rosen, H. (1999) Translational control of specific genes during differentiation of HL-60 cells. *J. Biol. Chem.* **274(20)**, 14,295–14,305.
3. Ashe, M. P., De Long, S. K., and Sachs, A. B. (2000) Glucose depletion rapidly inhibits translation initiation in yeast. *Mol. Biol. Cell.* **11(3)**, 833–848.
4. Meyuhas, O., Thompson, E. A. Jr., and Perry, R. P. (1987) Glucocorticoids selectively inhibit translation of ribosomal protein mRNAs in P1798 lymphosarcoma cells. *Mol. Cell. Biol.* **7(8)**, 2691–2699.
5. Tzamarias, D., Roussou, I., and Thireos, G. (1989) Coupling of GCN4 mRNA translational activation with decreased rates of polypeptide chain initiation. *Cell* **57(6)**, 947–954.
6. Kuhn, K. M., DeRisi, J. L., Brown, P. O., and Sarnow, P. (2001) Global and specific translational regulation in the genomic response of Saccharomyces cerevisiae to a rapid transfer from a fermentable to a nonfermentable carbon source. *Mol. Cell. Biol.* **21(3)**, 916–927.
7. Johannes, G., Carter, M. S., Eisen, M. B., Brown, P. O., and Sarnow, P. (1999) Identification of eukaryotic mRNAs that are translated at reduced cap binding complex eIF4F concentrations using a cDNA microarray. *Proc. Natl. Acad. Sci. USA* **96(23)**, 13,118–13,123.
8. Mikulits, W., Pradet-Balade, B., Habermann, B., Beug, H., Garcia-Sanz, J. A., and Mullner, E. W. (2000) Isolation of translationally controlled mRNAs by differential screening. *FASEB J.* **14(11)**, 1641–1652.
9. Zong, Q., Schummer, M., Hood, L., and Morris, D. R. (1999) Messenger RNA translation state: the second dimension of high- throughput expression screening. *Proc. Natl. Acad. Sci. USA* **96(19)**, 10,632–10,636.
10. Rickwood, D., ed. (1992) *Preparative Centrifugation: A Practical Approach.* IRL, Oxford.

7

Parallel Analysis of Gene Copy Number and Expression Using cDNA Microarrays

Jonathan R. Pollack

1. Introduction

A cDNA microarray consists of hundreds or thousands of polymerase chain reaction (PCR)–amplified cDNAs spotted onto a glass microscope slide, in a high-density pattern of rows and columns *(1)*. cDNA microarrays were first used widely to quantify gene expression across hundreds or thousands of genes simultaneously *(2,3)*. To measure gene expression, mRNAs from two different samples are differentially fluorescently labeled and cohybridized to a cDNA microarray, which is then scanned in dual wavelengths (**Fig. 1A**). For each cDNA element on the microarray, the ratio of fluorescence intensities reflects the relative abundance for that mRNA between the two samples. cDNA microarrays have been utilized to profile gene expression in a variety of organisms *(1,3–6)*. In humans, cDNA microarrays have been used extensively to characterize gene expression in cancer, in which patterns of gene expression reveal the molecular phenotypes of tumor cells *(7–9)*.

More recently, DNA microarrays have also been used to characterize alterations in genomic DNA copy number in cancer *(10–13)*. Alterations in DNA copy number, including the large chromosomal gains and losses that characterize aneuploidy, as well as more localized regions of gene amplification and deletion, are a near-universal finding in human cancer. Mapping chromosomal regions of DNA amplification and deletion is useful in the localization of oncogenes and tumor suppressor genes, respectively. Alterations in DNA copy number have been mapped genomewide using fluorescence *in situ* hybridization–based techniques, including comparative genomic hybridization

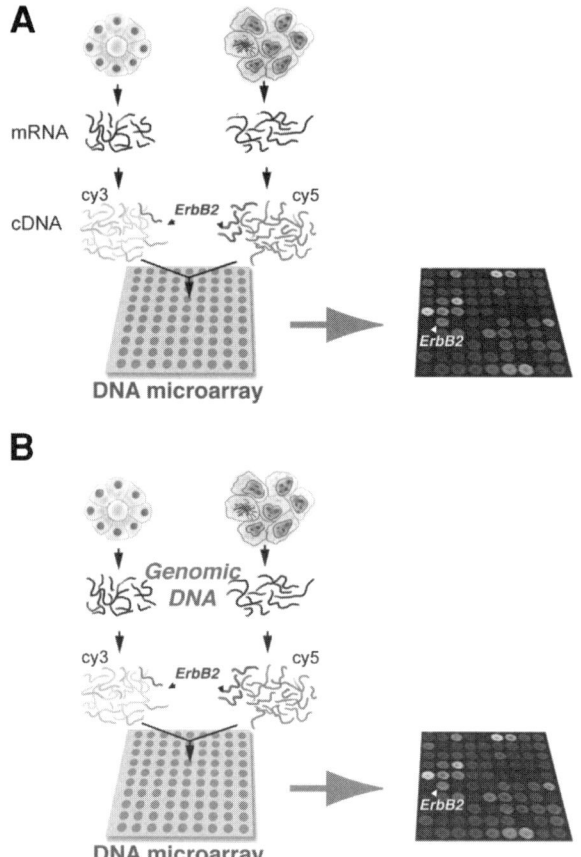

Fig. 1. Parallel analysis of gene expression and copy number using cDNA microarrays. (**A**) Measuring mRNA levels with cDNA microarrays. mRNAs from two different samples (normal and tumor shown) were differentially fluorescently labeled with Cy3 and Cy5, respectively, and cohybridized to a cDNA microarray, which was then scanned in dual wavelengths. For each cDNA element on the microarray, the ratio of fluorescence intensities reflects the relative abundance for that mRNA between the two samples. The ErbB2 gene is depicted as more highly expressed in the tumor sample. (**B**) Measuring gene copy number with cDNA microarrays. In this array CGH technique, genomic DNA from two different samples (normal and tumor shown) were differentially fluorescently labeled and cohybridized to a cDNA microarray. For each cDNA element on the microarray, the ratio of fluorescence intensities reflects the relative gene copy number (i.e., amplification or deletion) between the two samples. The ErbB2 gene is depicted as amplified in the tumor sample.

(CGH) *(14)* and spectral karyotyping *(15)*. However, these karyotype-based techniques have limited mapping resolution.

DNA microarrays provide a higher-resolution means to map alterations in DNA copy number *(10–13)*. cDNA microarrays in particular permit gene-by-gene analysis of alterations in DNA copy number *(12)*. To measure gene copy number using this array-based CGH (array CGH) technique, genomic DNAs from two different samples (typically tumor and normal) are differentially fluorescently labeled and cohybridized to a cDNA microarray, which is then scanned in dual wavelengths (**Fig. 1B**). For each cDNA element on the microarray, the ratio of fluorescence intensities reflects the relative gene copy number (i.e., amplification or deletion) between the two samples.

More important, cDNA microarrays now permit the parallel measurement of gene copy number and gene expression *(12)*. This is useful in the evaluation of candidate oncogenes. For example, only those genes within an amplicon that are highly expressed when amplified should be further considered as candidate oncogenes (**Fig. 2**). The parallel measurement of gene copy number and expression also permits genomic-scale analysis of the impact of widespread alterations in gene copy number on gene expression in cancer *(16)*. The methodology for labeling and hybridizing RNA to cDNA microarrays is now well established *(17)*. This chapter details the methodology for labeling and hybridizing genomic DNA to cDNA microarrays, requisite for the parallel analysis of gene copy number and expression.

2. Materials
2.1. Labeling of Genomic DNA
1. *Dpn*II restriction enzyme and 10X buffer (New England Biolabs).
2. PCR purification kit (Qiagen).
3. Bioprime labeling kit (Invitrogen).
4. 10X dNTP mix: 0.6 mM dCTP; and 1.2 mM each of dGTP, dATP, and dTTP in TE$_{8.0}$; prepare from ultrapure dNTP solutions (Pharmacia).
5. Cy5-dCTP, Cy3-dCTP (Amersham); supplied as 1 mM stock solutions.

2.2. Hybridization of Genomic DNA
1. Poly(dA-dT) (cat. no. P9764; Sigma, St. Louis, MO). Prepare a 5 µg/µL stock solution in TE$_{8.0}$.
2. Yeast tRNA (Invitrogen): Prepare a 5 µg/µL stock solution in TE$_{8.0}$.
3. Human Cot-1 DNA (Invitrogen); supplied as a 1 µg/µL stock solution.
4. 20X saline sodium citrate (SSC) solution.
5. 10% sodium dodecyl sulfate (SDS) solution.
6. Microcon 30 Filters (Amicon).
7. Hybridization chambers (Corning, Monterey Industries [www.montereyindustries.com], or equivalent).

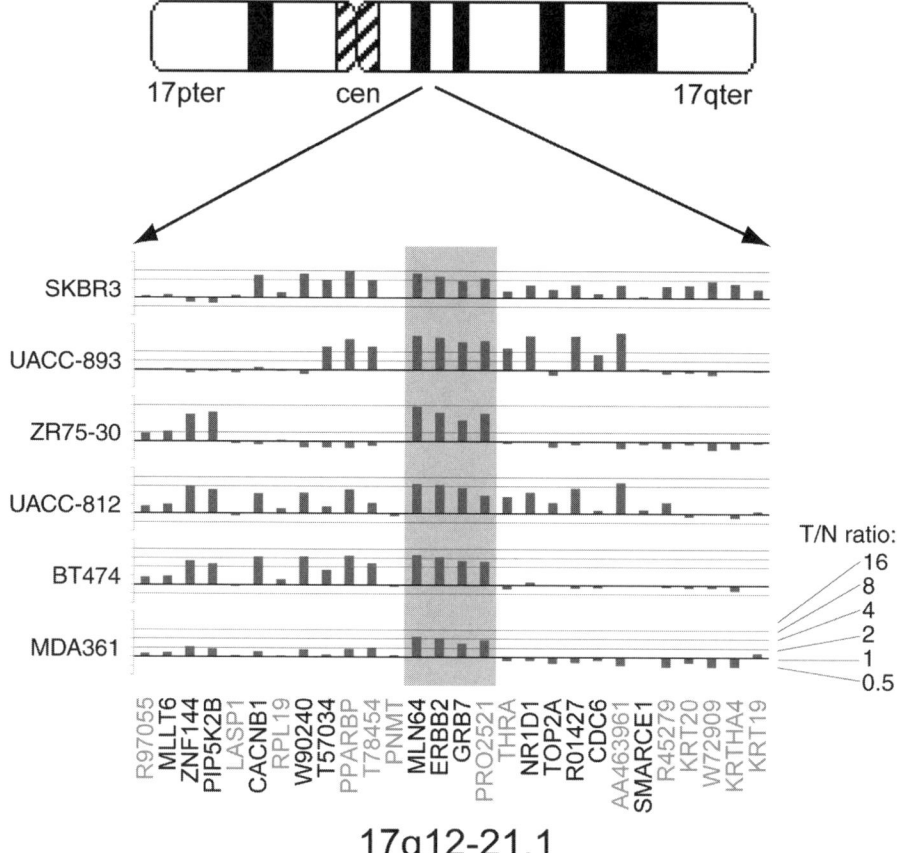

Fig. 2. cDNA microarray measurements of gene copy number across chromosomal region 17q12-21.1 in six different breast cancer cell lines. Fluorescence ratios (tumor/normal) are plotted on a $\log_2$ scale for genes ordered by nucleotide position (using the "Golden Path Human Genome Assembly"; http://genome.ucsc.edu) within this region of the chromosome. Fluorescence ratios >1 indicate gene amplification. Genes determined by parallel microarray analysis to be highly expressed when amplified are indicated by black text. The smallest region amplified among all samples is highlighted in gray and includes the highly expressed genes *MLN64*, *ERBB2*, and *GRB7*.

8. Glass microscope slide cover slips (22 × 60 mm) (Fisher).
9. Slide rack and three glass staining dishes (Wheaton).
10. Desktop centrifuge (Beckman Allegra 6R or equivalent), with microtiter plate carriers.

2.3. Data Reduction and Analysis

1. GenePix 4000B scanner (Axon), or equivalent, for imaging microarrays following hybridization.
2. GenePix software (Axon), or equivalent, for data reduction (e.g., matching pixels to cDNA spots, calculating fluorescence ratios).
3. Microsoft Excel, or various more specialized or custom microarray databases, for data manipulation.

3. Methods
3.1. cDNA Microarrays

Many universities, medical centers, and companies have established "core facilities" dedicated to printing cDNA microarrays. cDNA microarrays can also be purchased from various commercial vendors. cDNA microarrays printed on poly-L-lysine coated glass microscope slides produce consistently good hybridizations with low fluorescence backgrounds. Prior to use, cDNA microarrays should be processed to block nonspecific hybridization to the positively charged polylysine coating. Standard protocols exist for this procedure *(17)*.

3.2. Labeling and Hybridization of RNA

To measure gene expression, RNA should be isolated from cancer cell lines or tumors using Tri-reagent (MRC or Invitrogen), or equivalent. Standard protocols are available for the labeling and hybridization of RNA *(17)*.

3.3. Isolation of Genomic DNA

Genomic DNA can be isolated from cancer cell lines or tumors by any of several standard techniques (*see* **Note 1**). Anion-exchange column-based procedures (e.g., Genomic DNA Isolation Kit; Qiagen) offer a convenient option for preparation of high-quality genomic DNA. Tri-reagent (MRC) offers the possibility of isolating genomic DNA and RNA from the same tissue sample *(18)*. Genomic DNA concentration should be accurately quantified by ultraviolet spectrophotometry.

3.4. Labeling of Genomic DNA

1. In separate 1.5-mL Eppendorf tubes, digest 4 µg each of tumor and normal genomic DNA with *Dpn*II restriction enzyme (New England Biolabs), according to the manufacturer's instructions (*see* **Notes 2** and **3**).
2. Clean up the digested DNAs using the PCR purification kit (Qiagen), following the manufacturer's instructions. Resuspend each DNA sample in 21 µL of $TE_{8.0}$ in a 1.5-mL Eppendorf tube.

3. To each DNA sample, add 20 μL of 2.5X random primer mix (Bioprime labeling kit; Invitrogen) (*see* **Note 4**).
4. Boil for 5 min and then place on ice for 5 min.
5. To each tube on ice, add 5 μL of 10X dNTP mix (*see* **Note 5**), 3 μL of Cy5-dCTP (tumor DNA sample) or Cy3-dCTP (normal DNA sample), and 1 μL of concentrated Klenow enzyme (*see* **Note 6**).
6. Incubate at 37°C for 2 h.
7. Stop the reaction by adding 5 μL of stop solution (0.5 M $EDTA_{8.0}$; Bioprime labeling kit; Invitrogen).

3.5. Hybridization of Genomic DNA

1. Combine both labeled DNA samples together with 400 μL of $TE_{7.4}$ in a Microcon 30 filter, and spin at 12,000g for 10–12 min at room temperature in a microcentrifuge. The retained volume should be approx 20 μL (*see* **Note 7**). Discard the flowthrough.
2. Add an additional 450 μL of $TE_{7.4}$, spin at 12,000g for 10–12 min, and discard the flowthrough (*see* **Note 8**).
3. Add an additional 400 μL of $TE_{7.4}$, 4 μL of poly(dA-dT) stock solution, 20 μL of yeast tRNA stock solution, and 50 μL of human Cot-1 stock solution (*see* **Note 9**). Spin at 12,000g for 12 min and discard the flowthrough.
4. Invert the Microcon 30 filter into a new Eppendorf tube, and spin at 12,000g for 1 min to recover the labeled DNA.
5. Increase the volume of the recovered, labeled DNA mixture to 32 μL by adding ddH_2O.
6. Add 6.8 μL of 20X SSC and mix.
7. Add 1.2 μL of 10X SDS and mix gently to avoid forming bubbles.
8. Boil the hybridization mixture for 2 min, and then incubate at 37°C for 30 min.
9. Carefully pipet the hybridization mixture onto the cDNA microarray, and overlay with a 22 × 60 mm glass cover slip (*see* **Note 10**).
10. Incubate in a hybridization chamber at 65°C for 16 h.

3.6. Washing Microarrays After Hybridization

Following hybridization, wash the cDNA microarray to remove unbound labeled DNA. The following three sequential wash steps should be performed using a slide rack, transferring among three separate glass staining dishes, each containing volumes of 350–400 mL, with gentle agitation:

1. Wash 1: 2X SSC/0.03% SDS at 65°C for 5 min (*see* **Note 11**).
2. Wash 2: 1X SSC at room temperature for 5 min.
3. Wash 3: 0.2X SSC at room temperature for 5 min.

Following the third wash, the cDNA microarray should be spun dry using a desktop centrifuge at 50g and room temperature for 2 min.

3.7. Microarray Imaging, Data Reduction, and Analysis

Following hybridization, the cDNA microarray should be scanned in dual wavelengths using a GenePix 4000B (Axon) scanner, or equivalent. GenePix software, or equivalent, should be used to extract fluorescence intensity ratios for each cDNA element on the DNA microarray. To compensate for variable labeling efficiencies, fluorescence ratios should be normalized such that the average fluorescence ratio for all cDNA elements on the array is set to 1. Microsoft Excel, or any of several more specialized microarray databases, can be used to store and analyze microarray data. Finally, the Golden Path Human Genome Assembly (http://genome.ucsc.edu) can be used to map fluorescence ratios of cDNAs to chromosomal positions, in order to identify regions of DNA amplification and deletion (as in **Fig. 2**).

4. Notes

1. Microdissection techniques can be used to increase the cancer cell purity in tumor samples. DNA isolated from ethanol- or formalin-fixed, paraffin-embedded samples should be of sufficient quality for microarray hybridization.
2. Size reduction of genomic DNA serves to increase labeling efficiency.
3. Leukocytes from the whole blood of healthy donors are a convenient source for normal human genomic DNA.
4. The 2.5X random primer mix (Invitrogen) can also be prepared from separate components, to produce a final solution containing 125 mM Tris$_{6.8}$, 12.5 mM MgCl$_2$, 25 mM 2-mercaptoethanol, and 750 µg/mL of random octamers. The Bioprime labeling kit (Invitrogen) is a convenient and cost-efficient source of 2.5X random primer mix.
5. Be sure to use the 10X dNTP mix prepared in **Subheading 2.1., item 4**, and not the dNTP mix provided in the Bioprime labeling kit (Invitrogen), which contains biotin-dCTP.
6. The use of high-concentration Klenow (40–50 U/µL) is important for optimal labeling efficiency. The Bioprime labeling kit (Invitrogen) is a convenient and cost-efficient source, though other sources (e.g., New England Biolabs) are acceptable.
7. Centrifugation times are estimates. If necessary, here and in subsequent Microcon steps spin in additional 1-min increments until the volume of the retained solution is approx 20 µL.
8. This additional wash step serves to further remove unincorporated fluorescent nucleotides. *Important:* If labeling is successful, the retained labeled DNA mixture should appear bright purple.
9. Poly(dA-dT) and human Cot-1 serve to block undesirable hybridization to extended poly(A) tails and highly repetitive DNA, respectively, contained in a subset of cDNA microarray elements. Yeast tRNA functions to block nonspecific hybridization.

10. A total volume of 40 μL for the hybridization solution is appropriate when using a 22 × 60 mm cover slip. If using a different-sized cDNA microarray and cover slip, adjust the total volume of hybridization solution accordingly, while maintaining final SSC and SDS concentrations.
11. Performing the first wash at 65°C is important for optimizing the ratio of specific to nonspecific hybridization signal.

References

1. Shalon, D., Smith, S. J., and Brown, P. O. (1996) A DNA microarray system for analyzing complex DNA samples using two-color fluorescent probe hybridization. *Genome Res.* **6,** 639–645.
2. Schena, M., Shalon, D., Davis, R. W., and Brown, P. O. (1995) Quantitative monitoring of gene expression patterns with a complementary DNA microarray. *Science* **270,** 467–470.
3. DeRisi, J., Penland, L., Brown, P. O., et al. (1996) Use of a cDNA microarray to analyse gene expression patterns in human cancer. *Nat. Genet.* **14,** 457–460.
4. DeRisi, J. L., Iyer, V. R., and Brown, P. O. (1997) Exploring the metabolic and genetic control of gene expression on a genomic scale. *Science* **278,** 680–686.
5. White, K. P., et al. (1999) Microarray analysis of *Drosophila* development during metamorphosis. *Science* **286,** 2179–2184.
6. Jiang, M., Ryu, J., Kiraly, M., Duke, K., Reinke, V., and Kim, S. K. (2001) Genome-wide analysis of developmental and sex-regulated gene expression profiles in Caenorhabditis elegans. *Proc. Natl. Acad. Sci. USA* **98,** 218–223.
7. Ross, D. T., Scherf, U., Eisen, M. B., et al. (2000) Systematic variation in gene expression patterns in human cancer cell lines. *Nat. Genet.* **24,** 227–235.
8. Alizadeh, A. A., Eisen, M. M., Davis, R. E., et al. (2000) Distinct types of diffuse large B-cell lymphoma identified by gene expression profiling. *Nature* **403,** 503–511.
9. Perou, C. M., Sorlie, T., Eisen, M. B., et al. (2000) Molecular portraits of human breast tumours. *Nature* **406,** 747–752.
10. Solinas-Toldo, S., Lampel, S., Stilgenbauer, S., et al. (1997) Matrix-based comparative genomic hybridization: biochips to screen for genomic imbalances. *Genes Chromosomes Cancer* **20,** 399–407.
11. Pinkel, D., Segraves, R., Suder, D., et al. (1998) High resolution analysis of DNA copy number variation using comparative genomic hybridization to microarrays. *Nat. Genet.* **20,** 207–211.
12. Pollack, J. R., Perou, C. M., Alizadeh, A. A., et al. (1999) Genome-wide analysis of DNA copy-number changes using cDNA microarrays. *Nat. Genet.* **23,** 41–46.
13. Lucito, R., West, J., Reiner, A., et al. (2000) Detecting gene copy number fluctuations in tumor cells by microarray analysis of genomic representations. *Genome Res.* **10,** 1726–1736.
14. Kallioniemi, O. P., Kallioniemi, A., Sudar, D., et al. (1993) Comparative genomic hybridization: a rapid new method for detecting and mapping DNA amplification in tumors. *Semin. Cancer Biol.* **4,** 41–46.

15. Schrock, E., du Manoir, S., Veldman, T., et al. (1996) Multicolor spectral karyotyping of human chromosomes. *Science* **273,** 494–497.
16. Pollack, J. R., et al. (2002) Microarray analysis reveals a major direct role of DNA copy number alteration in the transcriptional program of human breast tumors. *Proc. Natl. Acad. Sci. USA*, **99,** 12963–12968.
17. Eisen, M. B. and Brown, P. O. (1999) DNA arrays for analysis of gene expression. *Methods Enzymol.* **303,** 179–205.
18. Chomczynski, P. (1993) A reagent for the single-step simultaneous isolation of RNA, DNA and proteins from cell and tissue samples. *Biotechniques* **15,** 532–534, 536, 537.

8

Genome-wide Mapping of Protein–DNA Interactions by Chromatin Immunoprecipitation and DNA Microarray Hybridization

Jason D. Lieb

1. Introduction

A critical part of understanding the mechanism and logic of cellular regulatory networks is understanding where enzymes and their regulatory proteins interact with the genome in vivo. From this, we can determine the genomic features that specify protein binding and simultaneously identify genes or other chromosomal elements whose function is affected by the binding. Recently, methods that combine well-established protocols for chromatin immunoprecipitations *(1–6)* with the surveying power of DNA microarrays have allowed researchers to create high-resolution, genomewide maps of the interaction between DNA-associated proteins and DNA *(7–9)*. Many variations of the method have been published, but all contain the same basic steps *(10)*: growth of cells, fixation, extract preparation, immunoprecipitation, fixation reversal, DNA purification, DNA amplification, microarray hybridization, and data analysis. The purpose here is to detail a single experimental method in yeast from start to finish, rather than to review all of the different protocols that have been used. The method described in this chapter worked for a particular set of DNA-associated proteins (Rap1p, Sir2p, Sir3p, and Sir4p), and their corresponding antibody–antigen interactions *(8)*. Since the strength, specificity, and mechanism of antibody–antigen and protein–DNA association vary widely, this protocol should be viewed as a starting point, rather than an absolute procedure.

The first step is to crosslink proteins at their sites of interaction with DNA. This is accomplished quickly and efficiently by adding formaldehyde directly

From: *Methods in Molecular Biology: vol. 224: Functional Genomics: Methods and Protocols*
Edited by: M. J. Brownstein and A. Khodursky © Humana Press Inc., Totowa, NJ

to living cells in culture. Crude extracts from these fixed cells are then prepared, sonicated to shear chromatin to an average size of approx 1 kb, and then used in immunoprecipitation reactions with antibodies raised against the DNA-associated protein of interest. DNA fragments enriched in each immunoprecipitation are then purified, amplified with a degenerate-primer polymerase chain reaction (PCR)–based method, fluorescently labeled, and hybridized to whole-genome DNA microarrays. These arrays contain spotted DNA fragments representing every open reading frame and intergenic region; in yeast this corresponds to roughly 13,000 unique genomic segments. The results of the hybridization allow one to identify which segments of the genome were enriched in the immunoprecipitation, and thereby to construct a genomewide map of in vivo protein–DNA interactions. The resolution of the method depends on the size of the sheared chromatin in the extract and the size of the arrayed segments of the genome. In contrast with other mapping methods, such as DNA footprinting, or chromatin immunoprecipitation followed by quantitative PCR, this procedure does not require any prior knowledge of a protein's binding targets and, by the same token, is not prejudiced by existing notions of where a given protein should bind.

2. Materials

1. YPD medium: In 950 mL of distilled water, add 10 g of yeast extract (cat. no. 0127-17-9; Difco, Detroit, MI) and 20 g of peptone (cat. no. 0118-17; Difco). Autoclave and add 50 mL of filter-sterile 40% dextrose. For plates, add 20 g of Difco agar (cat. no. 214010; Difco), autoclave, and gently stir on a hot plate until the medium is mixed well. Avoid causing bubbles. Pour 30–40 mL into a standard 100-mm plate.
2. 37% Formaldehyde (J.T. Baker).
3. Solution of 2.5 M glycine.
4. Cell-culture phosphate-buffered saline (PBS), pH 7.2: In 750 mL of ddH$_2$O, dissolve 8 g of NaCl, 0.2 g of KCl, 1.65 g of Na$_2$HPO$_4$, and 0.2 g of KH$_2$PO$_4$. Adjust the pH with 1 M HCl (usually add approx 1 mL). Bring the volume to 1 L with water. Autoclave.
5. Mini-BeadBeater-8^{TTT} (Biospec, www.biospec.com).
6. Glass beads (0.5 mm) (Biospec, www.biospec.com).
7. Screw-top tubes (2 mL), with rubber gasket.
8. Branson 250 microtip sonicator.
9. Syringe (6 mL) fitted with a 25-gage, 5/8-in. (16-mm) needle.
10. Stock solutions of protease inhibitors: 17.4 mg/mL of phenylmethylsulfonyl fluoride (PMSF) in ethanol or isopropanol; 100 mM benzamidine (cat. no. B6506; Sigma, St. Louis, MO) in water; 1 mg/mL of pepstatin in ethanol; 100 mM sodium metabisulfite in water. Aliquot and store all at –20°C up to 1 yr.

11. Beadbeater lysis buffer: 50 mM HEPES-KOH, pH 7.5; 10 mM MgCl$_2$; 150 mM KCl; 0.1 mM EDTA; 10% glycerol; 0.1% NP-40. This solution should be filter sterilized through a 0.22-μm filter and may be stored at room temperature. Just before use, add the following to the indicated concentration: 1 mM dithiothreitol (DTT), 1 mM sodium metabisulfate, 0.2 mM PMSF, 1 mM benzamidine (cat. no. B6506; Sigma), 1 μg/mL of pepstatin (see **Note 1**).
12. Immunoprecipitation buffer: 25 mM HEPES-KOH, pH 7.5; 150 mM KCl; 1 mM EDTA; 12.5 mM MgCl$_2$; 0.1% NP-40. Just before use, add 1 mM sodium metabisulfate, 0.2 mM PMSF, 1 mM benzamidine (cat. no. B6506; Sigma); 1 μg/mL of pepstatin (see **Note 1**).
13. Protein-G Sepharose (Amersham cat no. 17-0618-01).
14. Elution buffer: 50 mM Tris-HCl, pH 8.0, 10 mM EDTA, 1% sodium dodecyl sulfate (SDS).
15. 40 μM stock round A primer (name = T-PCRA); 5'-GTTTCCCAGTCACGAT CNNNNNNNN-3'.
16. 500 μM stock round B primer (name = T-PCRB): 5'-GTTTCCCAGTCAC GATC-3'.
17. Heated water bath (65°C).
18. Standard equipment for the production, hybridization, and analysis of DNA microarrays.

3. Methods
3.1. Culture, Crosslinking, and Preparation of Extract

Protein–DNA interactions may be sensitive to the physiological state of the cell. Therefore, it is important to consider the growth conditions of your culture carefully, and to control the growth conditions as tightly as possible. Consider the effects that any selection applied to the culture (nutrition, temperature, or drug) may have on the results. The following protocol describes the growth of wild-type yeast in rich medium to midexponential phase under standard laboratory conditions. Your experimental design may demand other conditions.

1. Streak the yeast onto YPD plates and incubate the plates at 30°C until single colonies appear. Pick a single colony and start a test tube culture in 5 mL of liquid YPD medium. Allow growth at 30°C overnight on a roller.
2. Dilute the overnight culture into 200 mL of YPD to an OD$_{600}$ of 0.05 or lower. Incubate at 30°C on a shaking platform. As a general rule, when diluting starter cultures, allow for three doublings so that the vast majority of cells have grown only at low density under the desired culture condition. A 200-mL culture grown to an OD$_{600}$ of 0.6 will yield enough cells for about six immunoprecipitation reactions.
3. When the culture reaches the desired density, add 37% formaldehyde solution (available from J.T. Baker) directly to the culture to a final concentration of 1%

formaldehyde. Continue incubation with shaking for 30 min at 30°C (*see* **Note 2**).
4. To quench the crosslinking reaction, add glycine to 125 m*M* (from a 2.5 *M* stock). Continue incubation with shaking for 10 min.
5. Pour the culture into four 50-mL polypropylene conical tubes, and pellet the cells at 1500*g* for 5 min. Pour off the supernatant, and wash each pellet by resuspension in 50 mL of cell culture PBS. Repellet the cells and wash an additional two times.
6. Resuspend each pellet in BeadBeater lysis buffer to a total volume of 3 mL. Combine all pellets in one new 15-mL polypropylene conical tube. Pellet the cells at 1500*g* for 5 min, and discard the supernatant.
7. Wash the combined pellet with 15 mL of BeadBeater lysis buffer. Pellet the cells at 1500*g* for 5 min, and discard the supernatant.
8. Weigh the pellet. Resuspend the cells such that 0.4 g of the cells is resuspended in a total volume of 1 mL of BeadBeater lysis buffer. For example, for a 1-g pellet the total resuspension volume is 2.5 mL. If your pellet is <0.4 g, resuspend it in 1 mL as well, but the pellet should weigh at least 0.1 g (*see* **Note 3**).
9. Pipet 1 mL of the resuspended cells into a 2-mL screw-top tube. Add 1 mL of 0.5-mm glass beads. Normal 1.5-mL snap-top tubes may be used instead, but the rubber-gasket screw tops prevent leaking. Place the tubes on ice for 5 min.
10. Lyse cells in the Mini-BeadBeater-8 with four 1-min sessions at the highest setting. Place the tubes on ice for 2 min between each session (*see* **Note 4**).
11. Recover the extract by pouring the bead/extract slurry into a 6-mL syringe fitted with a 25-gage, 5/8-in. (16-mm) needle. Allow the extract to drip into a clean 15-mL tube on ice. The plunger may be used to speed up the flow. After the last liquid has dripped out of the syringe, add 0.75 mL of BeadBeater lysis buffer to wash out the remaining extract. If the needle becomes clogged, insert the plunger, invert, and flick near the needle. The needle may need to be replaced if it remains clogged.
12. Keeping the extract in the 15-mL tube, sonicate for three 30-s sessions with a Branson 250 microtip sonicator at 50% duty cycle, power setting of 5. Ice for 2 min between sessions. Sonication further breaks the nuclei and should shear the DNA to a length of 300–1000 bp. Higher powers tend to cause frothing. The extract should be at about room temperature after each sonication session. If it becomes hot, decrease the time of each session and increase the number of sessions.
13. Spin the extract at full speed in a microfuge at 4°C for 5 min to clear the extract of debris and unlysed cells. Transfer the supernatant to a new tube and spin again to clear any remaining debris. Run 10 μL of the extract on an agarose gel. You should see a bright smear of DNA that ranges from 500–2000 bp, with a center near 1 kb. After this step, salt or detergent concentrations may be increased for more stringent immunoprecipitations. The extract may be frozen at –80°C, but I prefer to make fresh extract for each immunoprecipitation.

3.2. Chromatin Immunoprecipitation

1. Add the antibodies to the extract and incubate at 4°C for 4 h to overnight with nutation or rocking (*see* **Notes 5** and **6**). For each immunoprecipitation, I generally use 1–5 µg of antibodies and extract equivalent to approx 20 OD_{600} U of culture. For example, use a volume of extract equivalent to that derived from 20 mL of an OD_{600} 1 culture. The final volume of extract in one immunoprecipitation is typically 500 µL. Very rich extract may be diluted with BeadBeater lysis buffer to bring the volume up to 500 µL.
2. Prepare the protein-G sepharose beads (*see* **Note 7**). Place about 20 µL of dry protein-G sepharose beads in a clean 1.5-mL Eppendorf tube. Add 500 µL of immunoprecipitation buffer, and rock 10 min to wet the beads. Pellet the beads by centrifuging for 1 min at 4000 rpm in a microfuge (higher-speed spins may crush the beads). Aspirate the supernatant, and wash the beads again in immunoprecipitation buffer. Pellet the beads, and prepare a 50% (v/v) slurry of packed, swollen (wet) protein-G sepharose beads and immunoprecipitation buffer.
3. Recover the antibody–protein–DNA complexes. Add 50 µL of the protein-G sepharose bead slurry (equivalent to approx 25 µL of packed, swollen [wet] beads per immunoprecipitation). For pipeting, cut off the end of a pipet tip to make the opening wider. Incubate at 4°C for 1 to 2 h with nutation.
4. Wash the beads four times for 15 min each with 1 mL of immunoprecipitation buffer, spinning and aspirating as before. Do not take shortcuts here; more washes may increase the signal-to-noise ratio. Increase salt, detergent, or reducing agent concentrations if more stringent conditions are desired.
5. Elute the immunoprecipitated material from the beads by adding 100 µL of immunoprecipitation elution buffer to the washed beads. Mix and incubate at 65°C for 30 min. Spin down the beads and transfer 80 µL of the supernatant to a new tube.
6. Repeat **step 5** with 50 µL of immunoprecipitation elution buffer. Take 50 µL of the supernatant and pool the eluates.

3.3. Crosslink Reversal and DNA Purification

1. Incubate the eluate at 65°C for 6 h to overnight to reverse the crosslinks.
2. Add an equal volume of TE (pH 7.4), 1 µL of 20 mg/mL glycogen, and proteinase K to 100 µg/mL. Incubate at 37°C for 2 h.
3. Phenol/chloroform (1:1) extract, and transfer the aqueous phase to a new tube. Reextract the organic phase with 100 µL of TE, and pool the resulting aqueous phase with the aqueous phase from the previous step (*see* **Note 8**).
4. Add sodium acetate to 0.3 M. Add 2 vol of 100% ethanol. Place at –20°C for 1 h.
5. Pellet the DNA by centrifuging at full speed in a microfuge for 30 min. Wash the pellet with 70% ethanol, spin 10 min, and use a SpeedVac to dry. Do not overdry the pellet.

Table 1
Round A Reaction Mix Setup[a]

	Number of samples		
	1	3	5
5X buffer (μL)	1	3	5
dNTP mix (μL)	1.5	4.5	7.5
DTT (μL)	0.75	2.25	3.75
Bovine serum albumin (μL)	1.5	4.5	7.5
T7 Sequenase (μL)	0.3	0.9	1.5
Total (μL)	5.05	15.15	25.25
Stocks			
dNTPs	3 mM each		
DTT	0.1 M		
Bovine serum albumin	500 μg/mL		
T7 Sequenase	13 U/μL		

[a]Five microliters per reaction.

6. Resuspend the DNA in 25 μL of TE containing 100 μg/mL of RNase A (boiled, to preclude DNase activity). Vortex to dissolve any DNA on the sides of the tube. Incubate at 37°C for 30 min.

3.4. Amplification and Labeling of Immunoprecipitation-Enriched DNA

Round A consists of two rounds of DNA synthesis using template DNA from **Subheading 3.3., step 6**, a partially degenerate primer (primer A), and T7 Sequenase. Round B consists of 25 cycles of PCR using a primer (primer B) that anneals to the specific region of primer A. Round C fluorescently labels the amplified product with 25 cycles of PCR using primer B and Cy-dUTP (Pharmacia) (*see* **Note 9**).

1. Set up the round A reaction mix on ice as shown in **Table 1**. T7 Sequenase, 5X buffer, and Sequenase dilution buffer come together in a package from USB, and dNTPs are from Pharmacia (Ultrapure, 100 mM stocks):
2. In a PCR tube, set up the round A template mix as follows, and place it in the thermocycler: 7 μL of DNA to be amplified and labeled, 2 μL of 5X buffer, 1 μL of primer A (40 μM).
3. Program the thermocycler for the following cycle conditions: 2 min at 94°C, 2 min at 8°C, pause 2 min at 8°C (for the addition of 5 μL of reaction mix or 1 μL of enzyme), 8-min ramp to 37°C, 8-min hold at 37°C. For the 2-min pause in the first cycle, add 5 μL of reaction mix. In the second cycle, add 1 μL of T7

Table 2
Round B Reaction Mix Setup

	Samples		
	1	3	5
Template (diluted)	—	—	—
10X PCR buffer (μL)	10	30	50
100X dNTPs, 25 mM each (μL)	1	3	5
Primer B (μL)	2.5	7.5	12.5
Taq (μL)	1	3	5
25 mM MgCl$_2$ (μL)[a]	8	24	40
Water (μL)	62.5	187.5	312.5
Total (μL)	85	255	425
Stocks			
dNTPs[b]	25 mM each		
Primer B	500 pmol/μL		
Taq (Amplitaq)	5 U/μL		

[a][Mg^{2+}] = 2 mM final concentration.
[b]Final [dNTPs] = 250 μM.

    Sequenase enzyme mix. The mix for five reactions contains 3.5 μL of Sequenase dilution buffer (provided with Sequenase) and 1.5 μL of T7 Sequenase. Repeat once, for a total of two cycles.
4. After round A, dilute the product to 50 μL with 35 μL of 1X TE. Place 15 μL of the diluted template into a new PCR tube.
5. Set up the round B reaction mix on ice as shown in **Table 2**.
6. Add 85 μL of the mix to each of the tubes containing template.
7. Program the thermocycler as follows: 25 cycles of: 92°C for 30 s, 40°C for 30 s, 50°C for 30 s, and 72°C for 1 min.
8. Load the tubes into the thermocycler and start the program. When the cycling is complete, run 20 μL of the product on an agarose gel. A smear of DNA ranging from 100 bp to 1 kb, with a peak at about 350 bp, should be visible.
9. To fluorescently label the sample, transfer 15 μL of the round B product into a clean PCR tube, and set the tubes aside on ice.
10. Set up the round C (fluorescent labeling) reaction mix on ice as shown in **Table 3**.
11. Place 35 μL of the mix into each tube containing template. Place the tubes in the thermocycler, and run the following program: 25 cycles of: 92°C for 30 s, 40°C for 30 s, 50°C for 30 s, and 72°C for 2 min.

3.5. Purification and Hybridization of Probe
1. Place 450 μL of TE (pH 7.4) in a Microcon 30,000 mol wt cutoff spin filter (ym-30). Add the round C reactions to the filter, mixing thoroughly with the

Table 3
Round C Reaction Mix Setup

	Samples		
	1	3	5
Template	—	—	—
10X PCR buffer (provided with *Taq*)	5	15	25
F-100X dNTPs (see below)	0.5	1.5	2.5
Primer B	1	3	5
Cy dye	3	9	15
Taq	0.5	1.5	2.5
Water	21	63	105
25 mM MgCl$_2$	4	12	20
Total	35	105	175

F-100 dNTPs	
100 mM dATP	10 µL
100 mM dCTP	10 µL
100 mM dGTP	10 µL
100 mM dTTP	5 µL
1XTE	5 µL

TE. Centrifuge for 7 min at top speed. Inspect the filters to ensure that none have ruptured. Continue centrifuging in 30-s intervals until the volume is 5–10 µL.
2. Invert the filters into fresh tubes. Centrifuge for 1 min to harvest the labeled DNA. Labeled DNA may be stored at 4°C for hybridization the next day.
3. Mix the reference (generally Cy3-labeled) and experimental (generally Cy5-labeled) probes.
4. The final probe should be at 4X SSC, 0.2% SDS, containing competitor DNA as required. For a 22-mm^2 cover slip, the volume should be 15 µL; for a 22 × 40 mm cover slip, the volume should be 30 µL.
5. Set up the array in a hybridization chamber, placing 10 µL of 3X SSC on the edge of the slide to provide humidity.
6. Boil the probe for 2 min. Set aside several cover slips for the next step.
7. Pipet the probe onto the array, avoiding bubbles. Using forceps or another cover slip to help, immediately lay a cover slip over the array, avoiding bubbles. It is worth practicing this step several times with 10 µL of water and a blank microscope slide.
8. Close the hybridization chamber and submerge in a 65°C water bath. Hybridize for 4–24 h. I generally hybridize for 6 h. The quality of results from 6- and 15-h hybridizations is indistinguishable.

9. In separate glass slide chambers, prepare wash solutions no. 1 (1X SSC/0.03% SDS), no. 2 (0.2X SSC), and no. 3 (0.05X SSC).
10. Place the slide rack in wash no. 1. Disassemble the hybridization chamber and quickly place the array into the submerged slide rack. If the array is exposed to air while the cover slip starts to fall off, you may see high background fluorescent signal on the side of the array. Let the array sit in wash no. 1 until the cover slip slides off. Gently plunge the slide rack up and down several times to wash the array. Be sure not to scratch the array with the loose cover slip.
11. Manually transfer the array to the slide rack in wash no. 2. Plunge the slide rack up and down several times to wash the array.
12. Move the slide rack to wash no. 3 and rinse. Plunge the slide rack up and down several times to wash the array. It is critical to remove all the SDS.
13. To dry the slides, quickly transfer them to microtiter plate carriers with paper towels below the rack to absorb liquid, and centrifuge the slides for 5 min at $50g$.
14. Scan the array.

3.6. Considerations for Data Analysis

Microarray data from chromatin immunoprecipitation experiments require special treatment, but the best way to analyze the data is far from settled. Two steps in this method conspire to produce qualitative, rather than quantitative, results. First, the amplification step, while largely preserving the relative levels of DNA species in the original sample, introduces an element of variability that is specific to each DNA fragment in the immunoprecipitation-enriched fraction and the reference sample, and is therefore difficult to quantitate. Second, because the signal-to-noise levels vary with each immunoprecipitation experiment (i.e., how "clean" the immunoprecipitation is), and it is impossible to predict *a priori* how many DNA fragments a DNA-binding protein is supposed to bind in a particular experiment, it is extremely difficult to normalize ratio values across experiments. The general observation is that the raw ratio values can vary a great deal from experiment to experiment, but that the rank order of those ratios remains consistent (*see* **Note 9**).

Therefore, instead of using raw or normalized ratios, each arrayed genomic DNA segment is assigned a percentile rank based on the relative enrichment of the corresponding sequences in each immunoprecipitation. In my case, enrichment was measured relative to sheared genomic DNA that had been amplified using the same protocol as the immunoprecipitation samples (*see* **Note 9**). Each immunoprecipitation is repeated several times in parallel with a control immunoprecipitation. The median of the percentile rank values for each genomic segment is then calculated across all of the repeats for the control immunoprecipitations and the experimental immunoprecipitations, respectively. High median ranks result when particular genomic DNA fragments are enriched consistently. The idea is that a unique subclass of consistently enriched DNA fragments will appear in experimental immunoprecipitations, but not the control immunoprecipitations. This strategy seems to work well for DNA-associated proteins with many targets but may be more difficult to implement for proteins that bind to very few genomic loci. Alternate methods of data analysis *(7)* may be more appropriate in these cases.

Future efforts will be dedicated to making this experimental procedure more robust, and the analysis quantitative.

4. Notes

1. The salt concentration of this solution may not be ideal for your particular experiments. The optimal salt concentration should be determined in pilot experiments and can be adjusted conveniently by making one stock of buffer without KCl, and one stock with 1 M KCl, and mixing appropriately.
2. The ideal degree of crosslinking must be determined empirically for each protein-DNA-antibody combination. In some cases, no crosslinking may be required, whereas in others longer incubations are tolerated. The temperature of incubation during the crosslinking step can also be optimized.
3. In cases in which strains with different growth rates are being cultured for a single experiment, I synchronize the samples at this step by keeping the pellets from the faster-growing cultures on ice until the others have caught up. This did not affect my results, presumably because the yeast are dead and protein-DNA interactions are fixed. However, your proteins may behave differently. A 200-mL culture grown to an OD_{600} of 1 generally yields an approx 1-g pellet.
4. If a BeadBeater is not available, alternative methods of lysing the cells may presumably be used, but I have not tried any other method with this protocol.
5. Before adding antibodies, the extract may be precleared by incubating it with 1/10 vol of preswelled, washed protein-G beads for 1 h at 4°C. Spin down the beads at full speed in the microfuge, and transfer the supernatant to a new tube. I have not found that this step makes a discernible difference, but it may in individual cases.
6. Alternatively, the antibodies may be conjugated to protein-G beads prior to the immunoprecipitation. Add the antibodies to a slurry of approx 25 µL of swollen beads in a total volume of 200 µL of BeadBeater lysis buffer. Incubate for 45 min at 4°C. After incubation, discard the unbound antibodies and add the antibody-bead conjugates to the extract.
7. Optionally, block the protein-G beads to be used in the immunoprecipitation with bovine serum albumin (IgG free, 2 mg/mL; Sigma). I have not found that this step makes a discernible difference, but it may in individual cases.
8. The amount of DNA recovered at this step is very low (<10 ng). Alternate methods of DNA purification may be attempted (e.g., QiA-Quick columns), but the highest yield is probably obtained by simple extraction and precipitation.
9. As a reference sample for hybridization, I chose genomic DNA that had to be sheared to the same extent as DNA in the extracts, and then amplified. A large

pool of such amplified DNA was made initially and used as the reference for every sample. An alternative reference might be amplified DNA derived from each input extract. A third possibility is to use the DNA derived from mock or control immunoprecipitations as a reference, but since in the perfect case no DNA would be recovered in these experiments, I prefer to use one of the first two types of reference, and to hybridize the control samples to them on separate arrays.

Acknowledgments

The protocol described herein was based on the previous work of others in the chromatin and transcription fields and contains microarray-based protocols developed in the laboratory of Patrick O. Brown. I apologize to anyone whose work was not explicitly referenced. JDL was supported by the Helen Hay Whitney Foundation.

References

1. Braunstein, M., Rose, A. B., Holmes, S. G., Allis, C. D., and Broach, J. R. (1993) Transcriptional silencing in yeast is associated with reduced nucleosome acetylation. *Genes Dev.* **7(4),** 592–604.
2. Solomon, M. J., Larsen, P. L., and Varshavsky, A. (1988) Mapping protein-DNA interactions in vivo with formaldehyde: evidence that histone H4 is retained on a highly transcribed gene. *Cell* **53(6),** 937–947.
3. Strahl-Bolsinger, S., Hecht, A., Luo, K., and Grunstein, M. (1997) SIR2 and SIR4 interactions differ in core and extended telomeric heterochromatin in yeast. *Genes Devel.* **11(1),** 83–93.
4. Orlando, V. and Paro, R. (1993) Mapping polycomb-repressed domains in the bithorax complex using in vivo formaldehyde cross-linked chromatin. *Cell* **75(6),** 1187–1198.
5. Dedon, P. C., Soults, J. A., Allis, C. D., and Gorovsky, M. A. (1991) A simplified formaldehyde fixation and immunoprecipitation technique for studying protein-DNA interactions. *Anal. Biochem.* **197(1),** 83–90.
6. Carr, A. and Biggin, M. D. (1999) A comparison of in vivo and in vitro DNA-binding specificities suggests a new model for homeoprotein DNA binding in Drosophila embryos. *EMBO J.* **18(6),** 1598–1608.
7. Ren, B., Robert, F. Wyrick, J. J., et al. (2000) Genome-wide location and function of DNA binding proteins. *Science* **290(5500),** 2306–2309.
8. Lieb, J. D., Liu, X., Botstein, D., and Brown, P. O. (2001) Promoter-specific binding of Rap1 revealed by genome-wide maps of protein-DNA association. *Nat. Genet.* **28(4),** 327–334.
9. Iyer, V. A., Horak, C. A., Scafe, C. S., Botstein, D., Snyder, M., and Brown, P. O. (2001) Genomic binding distribution of the yeast cell-cycle transcription factors SBF and MBF. *Nature* **409,** 533–538.
10. Orlando, V. (2000) Mapping chromosomal proteins in vivo by formaldehyde-crosslinked chromatin immunoprecipitation. *Trends Biochem. Sci.* **25(3),** 99–104.

9

Statistical Issues in cDNA Microarray Data Analysis

Gordon K. Smyth, Yee Hwa Yang, and Terry Speed

1. Introduction

Statistical considerations are frequently to the fore in the analysis of microarray data, as researchers sift through massive amounts of data and adjust for various sources of variability in order to identify the important genes among the many that are measured. This chapter summarizes some of the issues involved and provides a brief review of the analysis tools that are available to researchers to deal with these issues.

Any microarray experiment involves several distinct stages. First, there is the design of the experiment. The researchers must decide which genes are to be printed on the arrays, which sources of RNA are to be hybridized to the arrays, and on how many arrays the hybridizations will be replicated. Second, after hybridization, there follow a number of data-cleaning steps or "low-level analysis" of the microarray data. The microarray images must be processed to acquire red and green foreground and background intensities for each spot. The acquired red/green ratios must be normalized to adjust for dye bias as well as for any systematic variation other than that owing to the differences among the RNA samples being studied. Third, the normalized ratios are analyzed by various graphic and numerical means to select differentially expressed genes or to find groups of genes whose expression profiles can reliably classify the different RNA sources into meaningful groups. The sections of this chapter correspond roughly to the various analysis steps.

The following notation is used throughout. The foreground red and green intensities are written Rf and Gf for each spot. The background intensities are Rb and Gb. The background-corrected intensities are R and G, where usually $R = Rf - Rb$ and $G = Gf - Gb$. The log differential expression ratio is

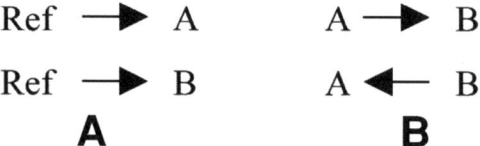

Fig. 1. Direct comparison (**B**) is more efficient than indirect comparison (**A**). Each arrow represents one microarray, the arrow by convention pointing toward the red labeled sample.

$M = \log_2 R/G$ for each spot. Finally, the log intensity of the spot is $A = 1/2 \log_2 RG$, a measure of the overall brightness of the spot. Note that the letter M is a mnemonic for minus, as $M = \log R - \log G$; while A is a mnemonic for add, as $A = (\log R + \log G)/2$. It is convenient to use base 2 logarithms for M and A so that M is in units of twofold change and A is in units of twofold increase in brightness. On this scale, $M = 0$ represents equal expression, $M = 1$ represents a twofold change among the RNA samples, $M = 2$ represents a fourfold change, and so on.

2. Experimental Design

Before carrying out a microarray experiment, one must decide how many microarray slides will be used and which mRNA samples will be hybridized to each slide. Certain decisions must be made in the preparation of the mRNA samples, such as whether the RNA from different animals will be pooled or kept separate and whether fluorescent labeling is to be done separately for each array or in one step for a batch of RNA. Careful attention to these issues will ensure that the best use is made of available resources, that obvious biases will be avoided, and that the primary questions of interest to the experimenter will be answerable. The literature on experimental design is still small. Kerr and Churchill *(1)* and Glonek and Solomon *(2)* apply ideas from optimal experimental designs to suggest efficient designs for some of the common microarray experiments. Pan et al. *(3)* consider sample size, and Speed and Yang *(4)* examine the efficiency of using a reference sample rather than direct comparison.

It is not possible to give universal recommendations appropriate for all situations, but the general principles of statistical experiment design apply to microarray experiments. In the simplest case in which the aim is to compare two mRNA samples, say *A* and *B*, it is virtually always more efficient to compare *A* and *B* directly by hybridizing them on the same arrays, rather than comparing them indirectly through a reference sample (**Fig. 1**) *(4)*. In an experiment in which the intention is to compare several mutant types with

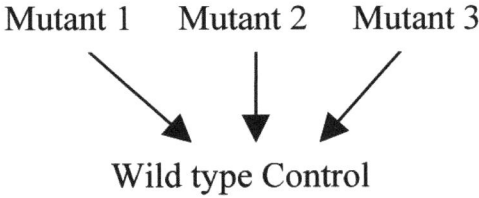

Fig. 2. WT RNA acts as a natural reference sample.

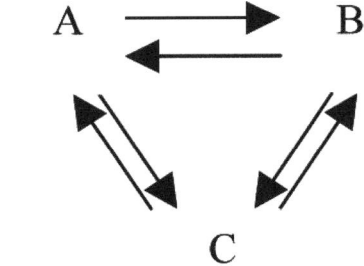

Fig. 3. Saturated design with dye-swap pairs.

Fig. 4. Possible design for a time-course experiment.

the wild type (WT), the obvious design treats the WT RNA effectively as a reference sample (**Fig. 2**). When more than two RNA samples are to be compared, and all comparisons are of interest, it may be appropriate to use a saturated design (**Fig. 3**). In time-course experiments, a loop design has been suggested (**Fig. 4**). For more complicated designs, with many samples to be compared, direct designs become more cumbersome, and it may be more appropriate to use a common reference sample. Factors to be considered in designing the experiment include the relative cost and availability of reference vs treatment RNA as well as the cost of the arrays themselves. In direct comparison experiments, it is generally advisable to use dye-swap pairs to minimize the effects of any gene-specific dye bias (**Fig. 3**).

The choice of experimental design depends not only on the number of different samples to be compared but on the aim of the experiment and on the comparisons that are of primary interest. For example, suppose the primary focus of an experiment involving a large series of tumor and normal tissues is on finding genes that are differentially expressed between the tumor and normal samples. Then direct tumor-normal comparisons on the same slide may be the best approach. By contrast, if the focus of the analysis is to determine tumor subtypes as in **ref. 5**, then the use of a common reference RNA on each array may be better. Here the choice follows from the aim of the study, although considerations of statistical efficiency also play a role. In the first case, tumor-normal comparisons could be made indirectly, via a common reference RNA, but precision would be lost in doing so.

3. Image Analysis

The primary purpose of the image analysis step is to extract numerical foreground and background intensities for the red and green channels for each spot on the microarray. The background intensities are used to correct the foreground intensities for local variation on the array surface, resulting in corrected red and green intensities for each spot, which become the primary data for subsequent analysis. A secondary purpose of the image analysis step is to collect quality measures for each spot that might be used to flag unreliable spots or arrays or to assess the reproducibility of each spot value.

The first step is to image the array using an optical scanner. The array is physically scanned to produce a digital record of the red and green fluorescence emissions at each point on the array. This digital record typically takes the form of a pair of 16-bit tiff images, one for each channel, which records the intensities at each of a large number of pixels covering the array. Depending on the scanner, a number of settings can be varied to improve the sensitivity of the resulting image, one of the most common being the photomultiplier tube (PMT) voltage. The PMT voltage is usually adjusted so that the brightest pixels are just below the level of saturation (2^{16}), thus increasing the sensitivity of the image analysis for the less bright pixels. Our own (unpublished) experiments with scanning a slide at varying PMT levels suggest that using different levels for the different channels has a negligible effect on the log ratios and ranks for the great majority of genes provided that an appropriate normalization method is used. In particular, any effect from varying the PMT levels is mitigated by using an intensity-based normalization method as described in **Subheading 4.**

The next step after scanning is to locate each spot on the slide. This is done mostly automatically by the image analysis software, using the known number and basic layout of spots on the slide, with some user intervention to increase reliability. Once a region containing a spot has been found, the image analysis software must segment the pixels into those in the spot itself (the foreground) and those in the background. There are a number of methods for doing this. The oldest method is the histogram method *(6)*. A mask is chosen surrounding each spot and a histogram is formed from the intensities of the pixels within the mask. Pixels are classified as foreground if their value is greater than a threshold and as background otherwise. Variations on this method are implemented in QuantArray software *(7)* for the GSI Lumonics scanner and in DeArray *(8)* by Scanalytics. The main advantage of this method is simplicity. However, the resulting foreground pixels are not necessarily connected, and the foreground and background intensities may be over and underestimated, respectively.

Other methods are designed to find spots as connected groups of foreground pixels. The simplest method is to fit a circle of constant diameter to all spots in the image. This is easy to implement and works nicely when all spots are circular and of the same size. In practice, this is not always the case. A generalization is to allow the circle's diameter to be estimated separately for each spot. GenePix *(9)* for the Axon scanner and Dapple *(10)* are two software programs that implement such algorithms. Dapple calculates the second differences (Laplacian) between the pixels in each small square and finds the brightest ring (circle) in the Laplacian images. Adaptive circle segmentation often works well, but spots are rarely perfectly circular, especially from noncommercial arrayers.

Two methods for segmentation that do not assume circularity of the spot are the watershed method *(11)* and seeded region growing *(12)*. Both methods require the specification of starting pixels or seeds. Adjoining pixels are then progressively added to the spot until adjacent spots appear to be distinctly less intense. Seeded region growing is implemented in the software Spot *(13)* and AlphaArray *(14)*. Both the watershed and seeded region growing methods allow for spots of general shapes.

Once the foreground pixels have been identified, the foreground intensity for the spot is usually estimated as the average intensity of all foreground pixels, since this should be directly proportional to the number of RNA molecules hybridized to the spot's DNA. When estimating the background intensity, it is more common to use the median intensity, but first there is a decision to be made regarding which pixels to include in the local background.

One choice for the local background is to consider all pixels that are outside the spot mask but within the bounding box. Such a method is implemented by ScanAlyze *(15)*. An alternative method used by QuantArray and ArrayVision *(16)* is to consider a disk between two concentric circles outside the spot mask. This method is in principle less sensitive to the performance of the segmentation procedure because the pixels immediately surrounding the spot are not used. Another method is to consider the valleys of the array, which are the background regions farthest from the nearest spot. The method is used by GenePix. It is also used by Spot as a quality control measure, although not for background correction. Since the valleys are farther from any spot than the other local background regions, the valley definition is less subject than the previous definitions to corruption by bright pixels affected by printed cDNA.

Any of the local background methods can result in background estimates that are higher than the foreground values either because of corruption by missegregated pixels or local artifacts or simply because of local variation.

The Spot software estimates the background using a nonlinear filter called morphological opening *(17)*. The filter has the effect of smoothing the entire slide image so that all local peaks, including artifacts such as dust particles as well as the spots themselves, are removed, leaving only the background intensities. Technically, the filter consists of a local minimum filter (erosion) followed by a local maximum filter (dilation). This method of background estimation has several advantages over the use of local background regions. First, it is less variable because the background estimates are based on a large window of pixel values and yet are not corrupted by bright pixels belonging to the actual spots. Second, it yields background intensity estimates at the actual spot location rather than merely nearby. Another characteristic is that the morphological background estimates are usually lower than the local background estimates and very rarely yield background estimates that are greater than the foreground values. Yang et al. *(18)* compared various segmentation and background estimation methods. They found that the choice of background method has a larger impact on the log ratios of intensities than the choice of segmentation method and that morphological opening provides a more reliable estimate of background than the other methods.

After having estimated the background intensities, it is almost universal practice to correct the foreground intensities by subtracting the background, $R = Rf - Rb$ and $G = Gf - Gb$. The adjusted intensities then form the primary data for all subsequent analyses. The motivation for background adjustment is the belief that a spot's measured intensity includes a contribution not specifically due to the hybridization of the target to the probe, such as nonspecific hybridization and fluorescence emitted from other chemicals on the glass. If such a contribution is present, one should measure and remove it to obtain a more accurate quantification of hybridization. An undesirable side effect of background correction is that negative intensities may be produced for some spots and hence missing values if log intensities are computed, resulting in loss of information associated with low channel intensities. Research has begun on more sophisticated methods of background adjustment that will produce positive adjusted intensities even when the background estimate happens to be larger than the foreground *(19)*. Empirical experience suggests that local background estimates often overestimate the true background while the morphological method may underestimate, and these differences have a marked impact on the M values for less intense spots. There is a need for further research on adaptive background correction methodologies that can produce intensities with consistent behavior regardless of the background estimator used.

4. Graphic Presentation of Slide Data

It is a good idea to use routinely a variety of exploratory graphic displays to examine the results of any microarray experiment. Graphic displays can help assess the success of the experiment, guide the choice of analysis tools, and highlight specific problems.

The first and most obvious diagnostic graphic is the well-known image in which the scanned microarray output images of the Cy3 and Cy5 channels are false-colored green and red, respectively, with yellow representing an equal balance of the two. Coregistration and overlay of the two channels offer a quick visualization of the experiment, revealing information on color balance, uniformity of hybridization, spot uniformity, background, and artifacts such as dust or scratches. Overlay images also provide a rough impression of the number of genes that are differentially expressed between the two samples.

Other diagnostic plots involve plotting the numerical values of the red and green intensities. Since the raw intensities are strictly positive and vary by orders of magnitude, they should almost always be log-transformed before plotting or carrying out further analysis. There are a number of reasons for this. First, the intensities in a successful microarray experiment typically span the full 16-bit range from 0 to 65,535, with the vast majority in the lower range of values, less than 1000. If the data are not transformed, the data must, by necessity, be presented in very compressed form in the low range. Calculating log values spreads the values more evenly across the range and provides readier visualization of the data. Second, the random variation, as measured by the standard deviation (SD) of the intensities, typically increases roughly linearly with the average signal strength. Converting to logarithms tends to make the variability more constant. Third, logarithms convert the ratios R/G to differences $M = \log R - \log G$.

Any negative values of R or G will have to be excluded from any analysis on the logarithmic scale. Negative values can be made very rare by using an unbiased background estimator as described in **Subheading 2**. In any case, spots with negative values for either R or G are usually too faint to show evidence of differential expression and therefore tend to be of less interest in any subsequent analysis.

The most common graphic display of data from a microarray slide is a scatter plot of the two channel intensities, $\log_2 R$ vs $\log_2 G$. Although such a plot is straightforward, the very high correlation between the two channel intensities always dominates the plot, making the more interesting features of

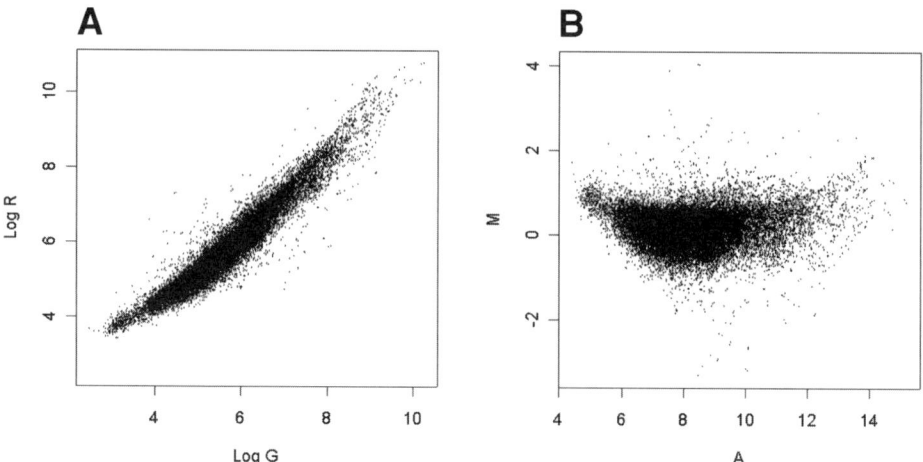

Fig. 5. (**A**) Scatter plot of log R vs log G and (**B**) *MA* plot. The central dip—an artifact—is more evident in (B) than in (A) and differentially expressed genes stand out more clearly. Data from the Nutt Lab, Walter and Eliza Hall Institute (WEHI).

the plot difficult to discern. Since the interest lies in deviations of the points from the diagonal line, it is beneficial to rotate the plot by 45° and to rescale the axes as in the *MA* plot of Dudoit et al. *(20)*, which has the *M* values on the vertical axis and the intensity *A* values on the horizontal axis. The *MA* plot serves to increase the room available to represent the range of differential expression and makes it easier to see nonlinear relationships between the log intensities (**Fig. 5**). It also displays the important relationship between differential expression and intensity, which is used in later analysis steps.

Box plots can be useful for comparing *M* values among various groups. A box plot displays graphically the so-called five-number summary of a set of numbers: the three quartiles and the maximum and minimum. The central box of the plot extends from the first to the third quartile and therefore encompasses the middle 50% of the data. **Figure 6** displays side-by-side box plots of the normalized *M* values for a series of six replicate arrays. The much longer box for array 5 shows that the interquartile range is much larger for this array. The different slides appear to be on varying scales, because of changes in PMT settings or other factors, and some rescaling seems to be called for to make the arrays more comparable.

A spatial plot of the background or *M* values can often reveal spatial trends or artifacts of various kinds. **Figure 7** shows a spatial plot of red channel morphological background for one array. Each spot on the array corresponds to

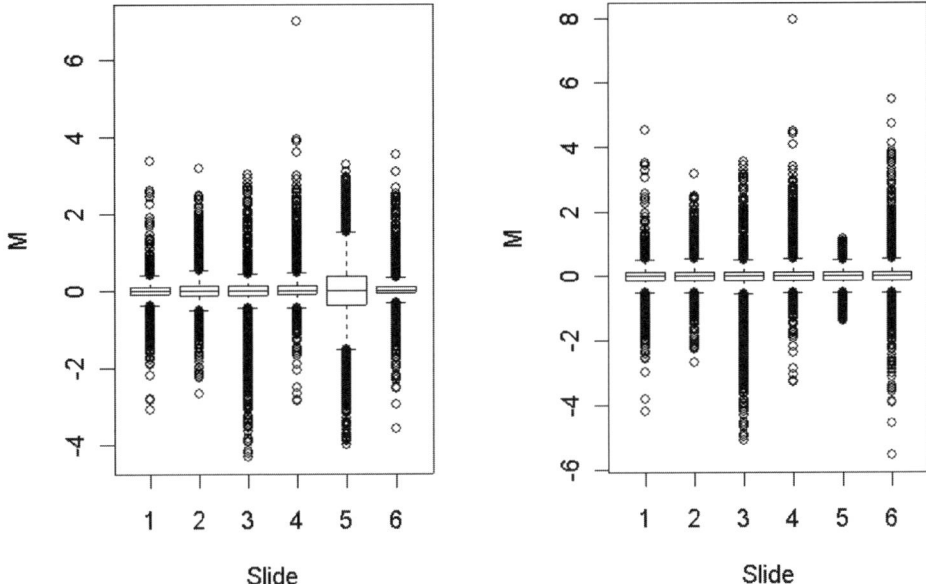

Fig. 6. (**A**) Side-by-side box plots of M-values from six arrays. The arrays are replicates except that three are dye-swap pairs of the others. Array 5 has a much larger spread than the others. (**B**) Box plots of the same arrays after scale normalization to equalize the median absolute deviation for each array. Data from the Corcoran Lab, WEHI.

Fig. 7. Spatial plot of morphological red channel background for a microarray slide. The gray scale goes from white for low background to black for high. The background is much higher around the edges and near the right edge. The array contains 19,200 spots in a 12×4 print-tip pattern. Data from the Scott Lab, WEHI.

one small square region on the plot. High background trends toward the edges of the plot stand out in the plot.

5. Normalization

The purpose of normalization is to adjust for any bias that arises from variation in the microarray technology rather than from biological differences among the RNA samples or the printed probes. Most common is red–green bias due to differences between the labeling efficiencies and scanning properties of the two fluors complicated perhaps by the use of different scanner settings. Other biases may arise from variation among spatial positions on a slide or between slides. Positions on a slide may vary because of differences among the print tips on the array printer, variation over the course of the print run, or nonuniformity in the hybridization. Differences among arrays may arise from differences in print quality or from differences in ambient conditions when the plates were processed. It is necessary to normalize the intensities before any subsequent analysis is carried out.

The need for normalization can be seen most clearly in self–self experiments, in which two identical mRNA samples are labeled with different dyes and hybridized to the same slide. Although there is no differential expression and one expects the red and green intensities to be equal, the red intensities often tend to be lower than the green intensities. Furthermore, the imbalance in the red and green intensities is usually not constant across the spots within and between arrays and can vary according to overall spot intensity, location on the array, slide origin, and possibly other variables.

Normalization can be carried out within each array or between arrays. The simplest and most widely used within-array normalization method assumes that the red–green bias is constant on the log scale across the array. The log ratios are corrected by subtracting a constant c to get normalized values $M = M - c$. The global constant c is usually estimated from the mean or median M value over a subset of the genes assumed to be not differentially expressed, but many other estimation methods have been proposed. Chen et al. *(6)* proposed iterative estimation of c as part of one of the first proposed normalization procedures. Kerr et al. *(21)* and Wolfinger et al. *(22)* have proposed the use of analysis of variance models for normalization. These methods are equivalent to subtracting a global constant as already discussed. Global normalization is still the most widely used in spite of evidence of spatial and intensity-dependent biases in numerous experiments. We favor more flexible normalization methods based on modern regression that take into account the effects of predictor variables such as spot intensity and location.

The next level of complication, which we have always found necessary, is to allow the correction c to vary between spots in an intensity-dependent

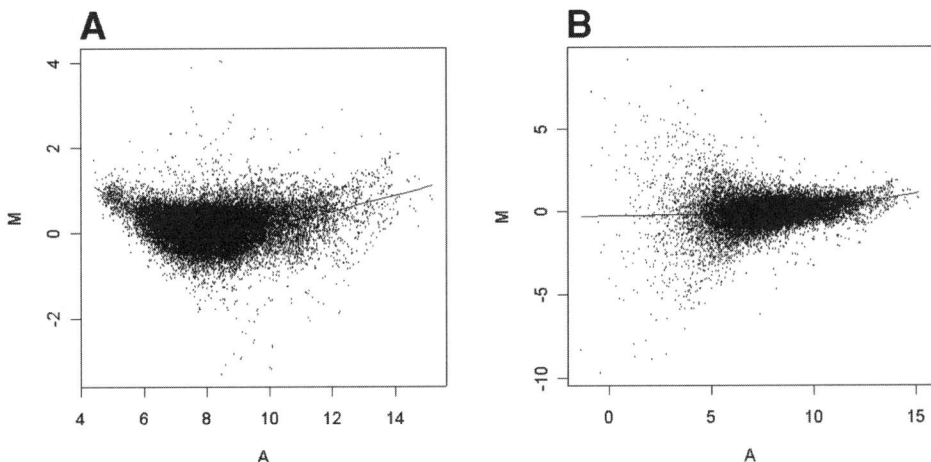

Fig. 8. Two *MA* plots of same microarray, (**A**) with morphological background and (**B**) with local median valley background. Data from the Nutt Lab, WEHI.

manner. In **Fig. 8**, a constant value for c would imply no trend between M and A. Instead, it can be seen that the majority of points lie on a curve, showing that the red–green bias depends on the intensity of the spot. Write $c(A)$ for the height of the curve at each value of A. We normalize the M values by subtracting this curve, $M = M - c(A)$. The curve is estimated using a suitable robust scatter plot smoother, such as local weighted regression (loess) *(23,24)*. A few other intensity-dependent methods have been proposed. Finkelstein et al. *(25)* proposed an iterative linear regression method that is essentially equivalent to what is known as robust linear regression in the statistical literature. This is similar to the aforementioned intensity-dependent normalization except that the curve $c(A)$ is constrained to be linear. Kepler et al. *(26)* proposed an intensity-dependent normalization that is similar to the loess method above but uses a different local regression method.

A further generalization is to use a different curve for different regions of the array, $M = M - c_i(A)$, in which i indexes the region of the array. We have found subarray normalization based on the print-tip groups to be particularly useful. Not only does this allow for physical differences among the actual tips of the printer head but the print-tip groups act as a surrogate for any spatial variation across the slide (**Fig. 9**).

There are often substantial scale differences among microarrays, because of changes in the PMT settings or other reasons. In these circumstances, we have also found it useful to scale-normalize among arrays, a simple scaling of the M values from a series of arrays so that each array has the same median absolute deviation (**Fig. 6**).

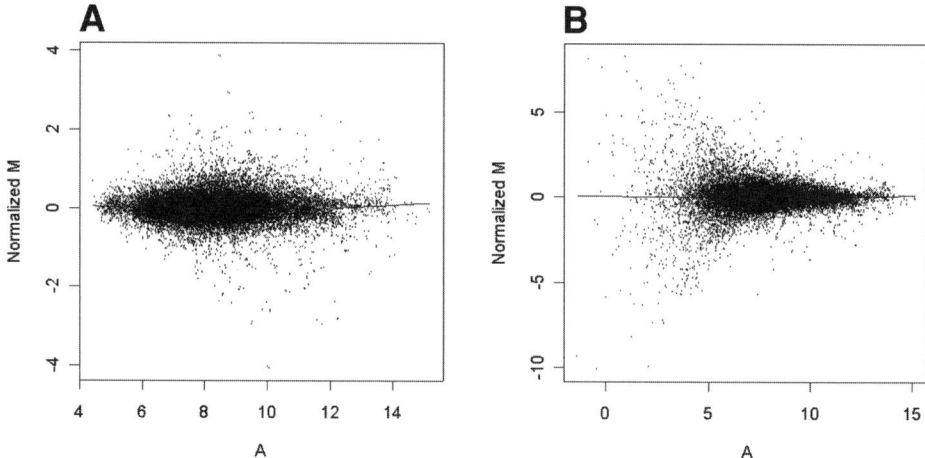

Fig. 9. Same two *MA* plots as in **Fig. 8** after print-tip loess normalization.

In all of the discussed normalization methods, it is usual to use all or most of the genes on the array. It can be useful to modify the normalization method if a suitable set of control spots is available. A traditional method is to use housekeeping genes for normalization. However, housekeeping genes often do show sample-specific bias. Housekeeping genes are also typically highly expressed, so they will not allow the estimation of dye biases for less expressed genes when the dye bias is intensity dependent. Housekeeping genes may also not be well represented on all parts of the plate, so spatial effects may not be well estimated. The most satisfactory set of controls is a specially designed microarray sample pool (MSP) titration series. MSP is analogous to genomic DNA as control with the exception that noncoding regions are removed. Typically, a concentration titration is done to span as wide an intensity range as possible. Theoretically all labeled cDNA sequences could hybridize to this mixed probe sample, so it should be minimally subject to any sample-specific biases. On the other hand, the use of all genes for normalization offers the most stability in terms of estimating spatial and intensity-dependent trends in the data. In some cases, it may be beneficial to use a compromise between the subarray loess curves and the global titration series curve *(24)*.

An alternative method is to select an invariant set of genes as described for oligonucleotide arrays by Schadt et al. *(27)* and Tseng et al. *(28)*. A set of genes is said to be invariant if their ranks are the same for both red and green intensities. In practice, the set of invariant or approximately invariant genes is too small for comprehensive normalization. When there are sufficient invariant genes, the use of invariant genes is similar to global intensity-dependent normalization as described above.

Subarray loess normalization is able to correct for a variety of spatial and intensity-dependent biases. It is advisable, however, using the exploratory plots mentioned in **Subheading 4.** to check whether other systematic effects exist in the data of which account should be made before the primary data analysis is carried out.

6. Quality Measures
6.1. Array Quality

It is important to assess the quality of the data obtained from each microarray experiment on a global array basis as well as on an individual spot basis. The quality of the results from each microarray will vary with cDNA purity, variations in the printing process, RNA quality, success in carrying out the hybridization protocols, and scanning effectiveness. A simple global assessment of quality is found in the distribution of log-intensity values in each of the two channels across the spots on the slide. Pixel intensities are usually scaled to be between 0 and 16 on the log base 2 scale. If the observed intensities fail to use the greater part of this scale, this is a strong indication that something is wrong; possibly the hybridization has failed. More precisely, we expect the intensity A values to span the majority of the response range. Control spots should be represented in this spread: null control spots such as blanks and printing buffers should have low intensities while housekeeping genes and titration series spots should show a range of higher intensities. At the same time, the intensity values should not be too dense around the largest value, suggesting that the scanner has been set too high and pixels have been saturated. This will lose discrimination and linearity of response on the log scale.

In most experiments, the great majority of genes should not be differentially expressed, so the range of M values for the bulk of genes should be much less than the range of A values. On an MA plot, the bulk of points should follow an elongated shape if the M and A axes are on a similar scale. If morphological background estimation has been used, the MA plot will typically follow an elongated comet shape with a long tail on the right. If a local background estimate has been used, the MA plot will typically follow a fan shape with, again, a long tail on the right (**Figs. 8** and **9**).

The ability of normalized intensities to follow a full range of values partly depends on the background level. A good-quality array will typically have relatively low background intensities and in particular a low average ratio of background to foreground intensity across the spots on the array.

The exploratory plots described in **Subheadings 4.** and **5.** will give an impression of array quality. The false-colored and spatial plots are particularly useful for judging spatial variation. Marked variation in the red–green dye bias across different parts of the array is an indication of quality problems. Although the subarray normalization will partly correct for spatial variation, strong variation will persist even after normalization and is an indication of problems with the experimental protocol.

6.2. Spot Quality

If the overall quality of an array is satisfactory, then it becomes relevant to assess the quality of individual spots. There are two broad approaches to this. The first is to assess the quality of a spot according to its physical characteristics. The second is to assess the quality of a spot according to whether the observed intensities for that spot are in general agreement with those from other spots printed with the same gene and hybridized with the same RNA. The first approach is an attempt to predict the repeatability of each spot's M value. Spots with low-quality scores are supposed to be less repeatable and are typically removed from subsequent analysis. The second is a data-based approach that observes repeatability empirically given a suitable series of replicate arrays or duplicate spots on the same array. A fully integrated approach to quality will include both approaches.

The first approach constructs quality measures for each spot from information collected by the image analysis program. Most image analysis programs routinely record a variety of spot details. These might include heterogeneity measures, such as SDs or interquartile ranges across pixels in the foreground and local background, as well as more basic details, such as spot area, perimeter, and location. Further quantities, such as circularity (area/perimeter^2) or interpixel coefficient of variation (CV) (SD/mean), can obviously be derived from the basic measures. In general, spots can be expected to be unreliable if they are very small or very large relative to the bulk of spots on the array, if they are markedly noncircular, if the background intensities are high, if the signal-to-noise ratio (SNR) is low, or if the foreground or background regions are very heterogeneous. Examples of such work can be found in **refs. *10*, *14*, *29*, and *30***. Buhler et al. *(10)* reject or accept spots based on brightness and position of the spot center. Brown et al. *(29)* consider pixel-level variability for each spot. Yang et al. *(30)* omit points with low intensities. Wang et al. *(14)* measure spot quality using a composite index involving spot size, SNR, level and heterogeneity of background, and saturation of pixels.

Examples of the more empirical quality approach can be found in **refs. *28* and *31***. Nadon et al. *(31)* reject spots that are judged to be outliers relative to a normal distribution for a series of M values from replicate slides. Tseng

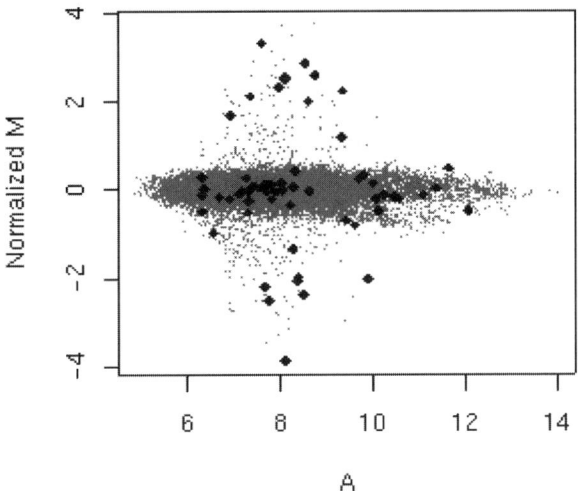

Fig. 10. Normalized *MA* plot for one microarray showing that very small spots are more variable than larger spots. Spots with areas <75 pixels are highlighted. Data from the Corcoran Lab, WEHI.

et al. *(28)* filter out genes according to the variability of duplicate spots on the same slide.

In all of these works, spots that are flagged as low quality are omitted from the primary analysis. Naturally this improves the look of the data, as indicated by a range of visual diagnostics. However, spots do not go from "good" to "bad" in a sharp way, and the cutoffs that are used to judge low quality are inevitably somewhat arbitrary. A more satisfactory approach would be to give less weight to lower-quality spots in a graduated way, with excellent spots getting full weight, down to excluding really bad spots entirely.

In the empirical quality approach, a more systematic approach to handling outlier spots can be achieved by using the robust estimation procedures discussed in **Subheading 7.** Robust estimators of location and scale will automatically downweight any *M* value that is discordant with other comparable values. Robust methods downweight outlying *M* values in a graduated way and avoid the need to choose an arbitrary cutoff.

In the physically based approach to quality, a graduated approach is more difficult. Ideally, quality measures should be found that predict the between-slide variance of the *M* values. Spots can then be weighted inversely according to the predicted variances. An obvious treatment of spot area, e.g., would be to weight small spots directly proportional to their area, such as $w = a/a_F$, in which w is the weight, a is the area of the spot in pixels, and a_F is the area of a full-sized spot. **Figure 10** demonstrates empirically that spots with small areas can be substantially more variable than larger spots. The correct treatment of other measures such as SNR is less obvious, although Brown et al. *(29)* and Wang et al. *(14)* have promising results. Wang et al. *(14)* demonstrated graphically an increasing trend relationship between spot variance and the spots' composite quality measure. However, we have not observed the same variance trends using data from our own institutions, and the results may be sensitive to the particular image analysis and background correction method that was used.

7. Selecting Differentially Expressed Genes
7.1. Ranking Genes

One of the core goals of microarray data analysis is to identify which of the genes show good evidence of being differentially expressed. This goal has two parts. The first is to select a statistic that will rank the genes in order of evidence for differential expression, from strongest to weakest evidence. The second is to choose a critical value for the ranking statistic above which any value is considered to be significant. The first goal is the more important of the two and, as it turns out, also the easier. The primary importance of ranking arises from the fact that only a limited number of genes can be followed up in a typical biological study. In many microarray studies, the aim is to identify a number of candidate genes for confirmation and further study. It will usually be practical to follow up only a limited number of genes, say 100, so it is most important to identify the 100 most likely candidates. The complete list of all genes that can be considered statistically significant may be of less interest if this list is too large to be followed up.

For simplicity, we assume in this section that we have data from the simplest possible experiment. We assume that we have a series of n replicate arrays on which samples A and B have been hybridized, and we wish to identify which genes are differentially expressed. Many data analysis programs sort the genes according to the absolute level of $\overline{M}$, in which $\overline{M}$ is the mean of the M values for any particular gene across the replicate arrays. This is known to be a poor choice because it does not take into account the variability of the expression levels for each gene *(32,33)*. The shortcoming of the method is that the variability of the M values over replicates is not constant across genes and genes with larger variances have a good chance of giving a large $\overline{M}$ statistic even if they are not differentially expressed. A better choice is to rank genes according to the absolute value of the t statistic:

$$t = \frac{\overline{M}}{s/\sqrt{n}}$$

Statistics in Microarray Analysis

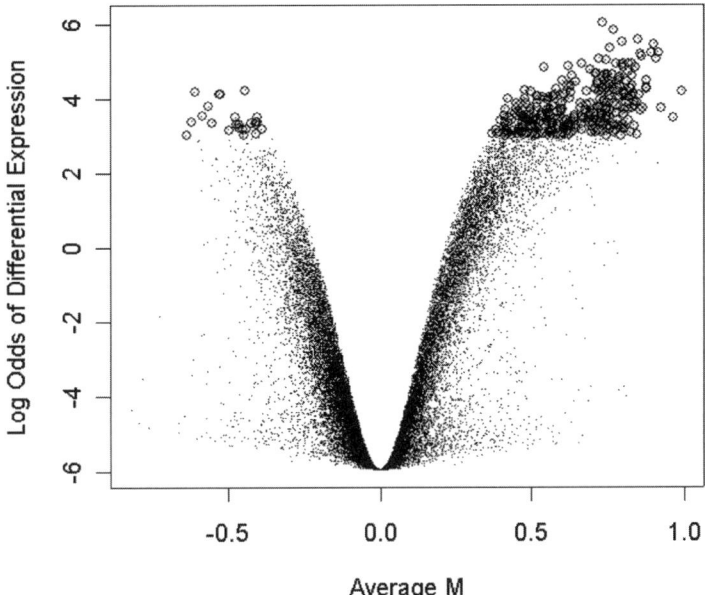

Fig. 11. Volcano style plot of the empirical Bayes B statistic for a series of six replicate arrays. Genes with log odds of differential expression greater than three have been highlighted for follow-up and confirmation. Data from the Corcoran Lab, WEHI.

in which s is the SD of the M values across the replicates for the gene in question, since this incorporates a different variability estimate for each gene. An added advantage of the t statistic is that it introduces some conservative protection against outlier M values and poor-quality spots. Any M value that is an outlier will give rise to a large SD s, which will usually prevent the gene in question from being spuriously identified as differentially expressed.

The ordinary t statistic is still not ideal because a large t statistic can be driven by an unrealistically small value for s. The shortcoming of the t statistic is the opposite of that of $\overline{M}$. Genes with small sample variances have a good a chance of giving a large t statistic even if they are not differentially expressed. A suitable compromise between the $\overline{M}$ and t statistics is therefore desirable.

Lönnstedt and Speed *(33)* adopt a parametric empirical Bayes approach to the problem of identifying differentially expressed genes. They produce a B statistic that is an estimate of the posterior log odds that each gene is differentially expressed (**Fig. 11**). Subject to the parametric assumptions being valid for the data, values for the B statistic greater than zero correspond to a greater than 50–50 chance that the gene in question is differentially expressed.

The B statistic is equivalent for the purpose of ranking genes to the penalized t statistic:

$$t = \frac{\overline{M}}{\sqrt{(a+s^2)/n}}$$

in which the penalty a is estimated from the mean and SD of the sample variances s^2. Tusher et al. *(32)* and Efron et al. *(34)* have used penalized t statistics of the form

$$t = \frac{\overline{M}}{(a+s)/\sqrt{n}}$$

when assessing differentially expressed genes for oligonucleotide microarrays. This differs slightly from the previous statistic in that the penalty is applied to the sample SD s rather than to the sample variance, s^2. Tusher et al. *(32)* choose a to minimize the coefficient of variation of the absolute t-values while Efron et al. [34] choose a to be the 90th percentile of the s values. These choices are driven by empirical rather than theoretical considerations. Efron et al. *(34)* use the above t value as the basis for a nonparametric empirical Bayes method leading to an estimated log odds that each gene is differentially expressed. Lönnstedt and Speed *(33)* show in a simulation that both forms of penalized t statistic are far superior to the mean $\overline{M}$ or to the ordinary t statistic for ranking differentially expressed genes.

The penalized t statistics can be extended in several natural ways to apply to more general experimental situations. If there are missing values for some arrays, perhaps because low-quality spots have been flagged for removal, then the value n in the denominator will reflect the actual number of observations for each gene rather than the total number of arrays.

The t statistic also extends naturally to more complicated experiment designs. For example, we might use a penalized two-sample t statistic if we are comparing samples A and B through a reference rather than directly on the same arrays. In that case, there will be n_A replicate arrays comparing sample A with reference RNA and n_B replicate arrays comparing B with the same reference and a two-sample t statistic:

$$t = \frac{\overline{M}_A - \overline{M}_B}{s_* \sqrt{\frac{1}{n_A} + \frac{1}{n_B}}}$$

in which $s_* = \sqrt{a + s^2}$ is the penalized pooled sample SD might be used. Here, $\overline{M}_A$ and $\overline{M}_B$ are the average of the M values for the two groups of arrays. For more complicated experimental designs, a multiple regression model will in general be estimated for each gene as, e.g., in **ref. 35**. In the general case, differential expression can be judged using a penalized t statistic of the form

$$t = \frac{b}{s_* \, \text{se}}$$

in which b is a regression coefficient estimated by the multiple regression that discriminates between the RNA samples of interest; se is the unscaled standard error for b returned by the multiple regression; and $s_* = \sqrt{a + s^2}$, in which s is the residual SD returned by the multiple regression. See Lönnstedt et al. *(36)*, who indicate the extension of the empirical Bayes B statistic to general experimental designs.

Another direction in which the t statistic can be generalized is to replace the sample mean $\overline{M}$ and sample SD s with location and scale estimators that are robust against outliers. This extension is very useful for microarray data because it is impossible to guarantee or adjust for the data quality of every individual spot. The general idea of robust estimation is to replace $\overline{M}$ and s with values that behave very much like $\overline{M}$ and s when the data actually are normally distributed but are insensitive to a small proportion of aberrant observations *(37) (38)*. For general microarray experiments, a robust multiple regression can be computed for each gene and a penalized t statistic formed from the robust versions of b, s, and se.

7.2. Assigning Significance

Having ranked the genes on the basis of a suitable statistic, the next step is to choose a cutoff value above which genes will be flagged as significant. The crux here is the need to control for the massive level of multiple testing inherent in the need to conduct a test for each gene.

A simple graphic method for assigning significance that is applicable even for single microarray experiments is to display the sorted genewise test statistics in a normal or t distribution probability plot. The bulk of the genes should follow an approximate straight line on the plot. Genes whose points deviate markedly from the line are identified by the method as significantly differentially expressed. Unfortunately, this remains an informal method because the implicit assumptions of normality for the M values and independence among genes are unlikely to be satisfied. The method tends in practice to overestimate the number of differentially expressed genes somewhat because the null distribution of the M values tends to have heavier tails than does the normal distribution. Tusher et al. *(32)* do use a variant of this method in conjunction with other multiple testing methods.

Shaffer *(39)* has reviewed the issues involved in multiple testing. The most stringent approach to multiple testing is to control for familywise error rate, which is the probability of at least one false positive among the genes selected as differentially expressed regardless of what configuration of the genes truly are differentially expressed. Dudoit et al. *(20)* consider a design for which two-sample *t* statistics are appropriate, comparing two RNA samples indirectly through a reference sample. They give a rigorous method for controlling the familywise error rate using a resampling method *(40)* that computes a step-down adjusted *p* value for each gene. Unfortunately, this method requires a moderate to large number of microarrays to give useful results. If, e.g., there are 15,000 distinct genes to be tested, then the method requires at least 16 microarrays to be able to detect differentially expressed genes because of the granularity of *p* values computed by resampling.

It can be argued that controlling the familywise error rate is unnecessarily stringent in the microarray testing context, because falsely selecting a handful of genes as differentially expressed will not be a serious problem if the majority of significant genes are correctly chosen. A less stringent and therefore more powerful method is to control the false discovery rate, defined to be the expected proportion of errors among the genes selected as significantly differentially expressed *(41)*. Tusher et al. *(32)*, Efron et al. *(34)* and Storey and Tibshirani *(42)* take an alternative approach to the false discovery rate. Rather than trying to control the false discovery rate, they treat it as an exploratory tool. After choosing the subset of differentially expressed genes by other means, they estimate the false discovery rate among this subset using a resampling method. Estimation of the false discovery rate, which is described in detail in **ref. 42**, relies formally on some assumptions about dependence among the genes that are difficult to verify in practice. However, this is a very promising approach.

The empirical Bayes methods of Efron et al. *(34)* and Lönnstedt and Speed *(33)* do not allow absolute cutoff values because the overall proportion of differentially expressed genes is an indeterminate parameter in the models. In using these methods, one has to specify a value in advance, say 1%, for the overall proportion of differentially expressed genes, including those that are detected and those that are not. Moving this value up or down will move all the posterior odds of differential expression up or down by a similar amount but will not change the order in which the genes are ranked. Efron et al. *(34)* suggest that the posterior odds can be calibrated post hoc by estimating the false discovery rate as discussed.

There exist other methods that can in principle give absolute cutoff values for differential expression *(6,22,43,44)*, in some cases even for as few as a single microarray in the experiment *(6,44)*. The price that is paid to achieve such results is that strong global distributional assumptions must be made about the red and green intensities. These assumptions are inevitably more simple than reality and seem to us too strong for routine data analysis.

8. Classification

Two very important uses for microarray data are to generate gene expression profiles that can (1) discriminate among different known cell types or conditions, e.g., between tumor and normal tissue or among tumors of different types, and (2) to identify different and previously unknown cell types or conditions, e.g., new subclasses of an existing class of tumors. The same problems arise when it is genes that are being classified: one might wish to assign an unknown cDNA sequence to one of a set of known gene classes, or one might wish to partition a set of genes into new functional classes on the basis of their expression patterns across a number of samples.

These dual tasks have been described as class prediction and class discovery in an influential article by Golub et al. *(45)*. In the machine-learning literature they are known as supervised and unsupervised learning, the learning in question being of the combinations of measurements—here gene expression values—that assign units to classes. In the statistical literature, they are known as discrimination and clustering. The distinction is important. Clustering or unsupervised methods are likely to be appropriate if classes do not exist in advance. If the classes are preexisting, then discriminant analysis or supervised learning methods are more appropriate and more efficient than clustering methods.

There are many powerful techniques for class prediction in the statistical and machine-learning literatures *(46–48)*. Such techniques invariably begin with data for which the existing class assignments are known, the so-called training set of units. These techniques can be effective even if some of the existing class assignments of the units are wrong or if there are unknown subclasses that would refine the existing classes. Indeed, there are well-established methods of evaluating the quality of prediction methods that, at the same time, check the assignment of individual units in the training set.

Cluster methods tend to be overused in microarray data analysis relative to discrimination methods. A common practice, e.g., is to suppress existing class assignments, use an unsupervised learning technique to define new classes and assign the units to these classes, and then see how well the existing class assignments are reflected in the new classes. A more direct and efficient approach would be to use a supervised method to discriminate the classes in conjunction with a method such as cross validation to evaluate the repeatability of the results on new data. The efficiency of direct discrimination over clustering becomes increasingly important as the prediction problem becomes more challenging.

Discrimination methods include linear discriminant analysis in various forms *(46)*, nearest-neighbor classifiers *(48)*, classification trees *(49)*, aggregating classifiers *(50–51)*, neural networks *(48)*, and support vector machines *(52–53)*. The first three methods are simple to apply once the genes have been filtered. The other methods are more sophisticated and require considerable skill in their application.

Dudoit et al. *(54)* compared the performance of different discrimination methods for the classification of tumors using gene expression data from three recent studies. Their main conclusion is that simple classifiers such as linear discrimination and nearest neighbors performed remarkably well compared to more sophisticated prediction methods such as aggregated classification trees.

There are factors other than accuracy that contribute to the merits of a given classifier. These include simplicity and insight gained into the predictive structure of the data. Linear discriminant methods are easy to implement and had low error rates in the study by Dudoit et al. *(54)* but ignore interactions among genes. Nearest-neighbor classifiers are simple, intuitive, and had low error rates compared with more sophisticated classifiers. While they are able to incorporate interactions among genes, they do so in a "black-box" way and give very little insight into the structure of the data. By contrast, classification trees are capable of exploiting and revealing interactions among genes. Trees are easy to interpret and yield information on the relationship between predictor variables and responses by performing stepwise variable selection. However, classification trees tend to be unstable and lacking in accuracy. Their accuracy can be greatly improved by aggregation (bagging or boosting). As more data become available, one can expect to observe an improvement in the performance of aggregated classifiers relative to simpler classifiers, because trees should be able to correctly identify interactions.

The use of clustering methods to identify group coregulated genes is an area of very active research, stimulated by influential articles such as those by Eisen et al. *(55)* and Alizadeh et al. *(5)*. The most popular clustering methods are nicely reviewed by Quackenbush *(53)*. Recent work includes that of Hastie et al. *(56)*, who formed clusters around the largest principal components of the data; Lazzeroni and Owen *(57)*, who proposed models in which each gene can belong to more than one cluster to none at all as different characteristics are considered; Parmigiani et al. *(58)*, who considered more general probabilistic models; and Lin et al. *(35)*, who clustered genes on the basis of regression coefficients estimated by a linear model.

9. Conclusion

Attention to statistical issues at each stage of microarray data analysis can ensure that the best use is made of available resources, that biases of various sorts are avoided, and that reliable conclusions are made. Software to carry out the analyses discussed in this chapter is described by Dudoit et al. *(59)* and Dudoit and Yang *(60)*.

Acknowledgments

Thanks to Matthew Ritchie and Natalie Thorne for discussions and to Matthew Ritchie for **Fig. 10**. We are also grateful to Drs. Lynn Corcoran, Stephen Nutt, and Hamish Scott for permission to use data from their laboratories at WEHI.

References

1. Kerr, M. K. and Churchill, G. A. (2001) Experimental design for gene expression microarrays. *Biostatistics* **2,** 183–201.
2. Glonek, G. F. V. and Solomon, P. J. (2002) Factorial designs for microarray experiments. Technical Report, Department of Applied Mathematics, University of Adelaide, Australia.
3. Pan, W., Lin, J., and Le, C. (2002) How many replicates of arrays are required to detect gene expression changes in microarray experiments? A mixture model approach. *Genome Biol.* **3(5),** research0022.1–0022.10.
4. Speed, T. P. and Yang, Y. H. (2002) Direct versus indirect designs for cDNA microarray experiments. Technical Report 616, Department of Statistics, University of California, Berkeley.
5. Alizadeh, A. A, Eisen, M. B., Davis, R. E., et al. (2000) Distinct types of diffuse large B-cell lymphoma identified by gene expression profiling. *Nature* **403(6769),** 503–511.
6. Chen, Y., Dougherty, E. R., and Bittner, M. L. (1997) Ratio based decisions and the quantitative analysis of cDNA microarray images. *J. Biomed. Opt.* **2,** 364–374.
7. QuantArray Analysis Software. http://lifesciences.perkinelmer.com.
8. Scanalytics MicroArray Suite. www.scanalytics.com.
9. GenePix Pro microarray and array analysis software, Axon Instruments www.axon.com.
10. Buhler, J., Ideker, T., and Haynor, D. (2000) Dapple: improved techniques for finding spots on DNA microarrays. CSE Technical Report UWTR 2000-08-05, University of Washington.
11. Beucher, S. and Meyer, F. (1993) The morphological approach to segmentation: the watershed transformation: mathematical morphology in image processing. *Opt. Eng.* **34,** 433–481.
12. Adams, R. and Bischof, L. (1994) Seeded region growing. *IEEE Trans. Pattern Anal. Machine Intelligence* **16,** 641–647.

13. Buckley, M. J. (2000) *Spot User's Guide*, CSIRO Mathematical and Information Sciences, Sydney, Australia. www.cmis.csiro.au/iap/Spot/spotmanual.htm.
14. Wang, X., Ghosh, S., and Guo, S.-W. (2001) Quantitative quality control in microarray image processing and data acquisition. *Nucleic Acids Res.* **29(15)**, E75–5.
15. Eisen, M. B. (1999) *ScanAlyze User Manual*, Stanford University, Palo Alto. http://rana.lbl.gov.
16. ArrayVision, Imaging Research. http://imaging.brocku.ca.
17. Soille, P. (1999) *Morphological Image Analysis: Principles and Applications*, Springer, New York.
18. Yang, Y. H., Buckley, M. J., Dudoit, S., and Speed, T. P. (2002) Comparison of methods for image analysis on cDNA microarray data. *J. Computat. Graph. Stat.* **11**, 108–136.
19. Kooperberg, C., Fazzio, T. G., Delrow, J. J., and Tsukiyama, T. (2002) Improved background correction for spotted cDNA microarrays. *J. Computat. Biol.* **9**, 55–66.
20. Dudoit, S., Yang, Y. H., Speed, T. P., and Callow, M. J. (2002) Statistical methods for identifying differentially expressed genes in replicated cDNA microarray experiments. *Statistica Sinica* **12**, 111–140.
21. Kerr, M. K., Martin, M., and Churchill, G. A. (2000) Analysis of variance for gene expression microarray data. *J. Computat. Biol.* **7**, 819–837.
22. Wolfinger, R. D., Gibson, G., Wolfinger, E. D., Bennett, L., Hamadeh, H., Bushel, P., Afshari, C., and Paules, R. S. (2001) Assessing gene significance from cDNA microarray expression data via mixed models. *J. Computat. Biol.* **8**, 625–637.
23. Yang, Y. H., Dudoit, S., Luu, P., and Speed, T. P. (2001) Normalization for cDNA microarray data, in *Microarrays: Optical Technologies and Informatics* (Bittner, M. L. Chen, Y. Dorsel, A. N. and Dougherty, E. R., eds.), Proceedings of SPIE, vol. 4266.
24. Yang, Y. H., Dudoit, S., Luu, P., Lin, D. M., Peng, V., Ngai, J., and Speed, T. P. (2002) Normalization for cDNA microarray data: a robust composite method addressing single and multiple slide systematic variation. *Nucleic Acids Res.* **30(4)**, E15.
25. Finkelstein, D. B., Gollub, J., Ewing, R., Sterky, F., Somerville, S., and Cherry, J. M. (2001) Iterative linear regression by sector, in *Methods of Microarray Data Analysis. Papers from CAMDA 2000.* (Lin S. M. and Johnson, K. F., eds.) Kluwer Academic, pp. 57–68.
26. Kepler, T. B., Crosby, L., and Morgan, K. T. (2000) Normalization and analysis of DNA microarray data by self-consistency and local regression, Santa Fe Institute Working Paper, Santa Fe, NM.
27. Schadt, E. E., Li, C., Ellis, B., and Wong, W. H. (2002) Feature extraction and normalization algorithms for high-density oligonucleotide gene expression array data. *J. Cell. Biochem.* **84(Suppl. 37)**, 120–125.
28. Tseng, G. C., Oh, M.-K., Rohlin, L., Liao, J. C., and Wong, W. H. (2001) Issues in cDNA microarray analysis: quality filtering, channel normalization, models of variations and assessment of gene effects. *Nucleic Acids Res.* **29**, 2549–2557.

29. Brown, C. S., Goodwin, P. C., and Sorger, P. K. (2000) Image metrics in the statistical analysis of DNA microarray data. *Proc. Natl. Acad. Sci. USA* **98,** 8944–8949.
30. Yang, M. C., Ruan, Q.-G., Yang, J. J., Eckenrode, S., Wu, S., McIndoe, R. A., and She, J.-X. (2001) A statistical procedure for flagging weak spots greatly improves normalization and ratio estimates in microarray experiments. *Physiol. Genomics* **7,** 45–53.
31. Nadon, R., Shi, P., Skandalis, A., Woody, E., Hubschle, H., Susko, E., Rghei, N., and Ramm, P. (2001) Statistical methods for gene expression arrays, in *Microarrays: Optical Technologies and Informatics* Proceedings of SPIE, vol. 4266, (Bittner, M. L., Chen, Y., Dorsel, A. N., and Dougherty, E. R. eds.), pp. 46–55.
32. Tusher, V., Tibshirani, R., and Chu, G. (2001) Significance analysis of microarrays applied to the ionizing radiation response. *Proc. Natl. Acad. Sci. USA* **98,** 5116–5124.
33. Lönnstedt, I. and Speed, T. P. (2002) Replicated microarray data. *Statistica Sinica* **12,** 31–46.
34. Efron B., Tibshirani, R., Storey J. D., and Tusher V. (2001) Empirical Bayes analysis of a microarray experiment. *J. Am. Stat. Assoc.* **96,** 1151–1160.
35. Lin, D. M., Yang, Y. H., Scolnick, J. A., Brunet, L. J., Peng, V., Speed, T. P., and Ngai, J. (2002) A spatial map of gene expression in the olfactory bulb, Department of Molecular and Cell Biology, University of California, Berkeley.
36. Lönnstedt, I., Grant, S., Begley, G., and Speed, T. P. (2001) Microarray analysis of two interacting treatments: a linear model and trends in expression over time. Technical Report, Department of Mathematics, Uppsala University, Sweden.
37. Huber, P. J. (1981) *Robust Statistics*, Wiley, New York.
38. Marazzi, A. (1993) *Algorithms, Routines and S Functions for Robust Statistics*, Wadsworth & Brooks/Cole, CA.
39. Shaffer, J. P. (1995) Multiple hypothesis testing. *Annu. Rev. Psychol.* **46,** 561–576.
40. Westfall, P. H. and Young, S. S. (1993) *Re-Sampling Based Multiple Testing*, Wiley, New York.
41. Benjamini, Y. and Hochberg, Y. (1995) Controlling the false discovery rate: a practical and powerful approach to multiple testing. *J. Roy. Stat. Soc. Ser.* **57,** 289–300.
42. Storey, J. D. and Tibshirani, R. (2001) Estimating false discovery rates under dependence with applications to DNA microarrays, Technical Report, Department of Statistics, Stanford University.
43. Ideker, T., Thorsson, V., Siegel, A. F., and Hood, L. (2000) Testing for differentially-expressed genes by maximum-likelihood analysis of microarray data. *J. Computat. Biol.* **7(6),** 805–817.
44. Newton, M. A., Kenziorski, C. M., Richmond, C. S., Blattner, F. R., and Tsui, K. W. (2001) On differential variability of expression ratios: improving statistical inference about gene expression changes from microarray data. *J. Computat. Biol.* **8,** 37–52.

45. Golub, T. R., Slonim, D. K., Tamayo, P., Huard, C., Gaasenbeek, M., Mesirov, J. P., Coller, H., Loh, M. L., Downing, J. R., Caligiuri, M. A., Bloomfield, C. D., and Lander, E. S. (1999) Molecular classification of cancer: class discovery and class prediction by gene expression monitoring. *Science* **286,** 531–537.
46. Mardia, K. V., Kent, J. T., and Bibby, J. M. (1979) *Multivariate Analysis*, Academic, London.
47. McLachlan, G. J. (1992) *Discriminant Analysis and Statistical Pattern Recognition*, Wiley, New York.
48. Riply, B. D. (1996) *Pattern Recognition and Neural Networks*, Cambridge University Press, Cambridge.
49. Breiman, L., Friedman, J. H., Olsen, R. A., and Stone, C. J. (1984) *Classification and Regression Trees*, Wadsworth, Monterey, CA.
50. Breiman, L. (1996) Bagging predictors. *Machine Learning* **24,** 123–140.
51. Breiman, L. (1998) Arcing classifiers. *Ann. Stat.* **26,** 801–824.
52. Brown, M. P., Grundy, W. N., Lin, D., Cristianini, N., Sugnet, C. W., Furey, T. S., Ares, M. Jr., and Haussler, D. (2000) Knowledge-based analysis of microarray gene expression data by using support vector machines. *Proc. Natl. Acad. Sci. USA* **97,** 262–267.
53. Quackenbush, J. (2001) Computational analysis of microarray data. *Nat. Rev. Genet.* **2,** 418–427.
54. Dudoit, S., Fridlyand, J., and Speed, T. P. (2002) Comparison of discrimination methods for the classification of tumors using gene expression data. *J. Am. Stat. Assoc.* **97,** 77–87.
55. Eisen, M. B., Spellman, P. T., Brown, P. O., Botstein, D. (1998) Cluster analysis and display of genome-wide expression patterns. *Proc. Natl. Acad. Sci. USA* **95,** 14,863–14,868.
56. Hastie, T., Tibshirani, R., Eisen, M. B., Alizadeh, A., Levy, R., Staudt, L., Chan, W. C., Botstein, D., and Brown, P. (2000) "Gene shaving" as a method for identifying distinct sets of genes with similar expression patterns. *Genome Biol.* **1(2),** 0003.1–0003.21.
57. Lazzeroni, L. and Owen, A. B. (2002) Plaid models for gene expression data. *Statistica Sinica* **12,** 61–86.
58. Parmigiani, G., Garrett, E. S., Anbazhagan, R., and Gabrielson, E. (2002) A statistical framework for expression-based molecular classification in cancer, Technical Report, Department of Biostatistics, Johns Hopkins University.
59. Dudoit, S., Yang, Y. H., and Bolstad, B. (2002) Using R for the analysis of DNA microarray data. *R News* **2(1),** 24–32.
60. Dudoit, S. and Yang, Y. H. (2003) Bioconductor R packages for exploratory analysis and normalization of cDNA microarray data, in *The Analysis of Gene Expression Data: Methods and Software* (Parmigiani, G., Garrett, E. S., Irizarry, R. A., and Zeger, S. L., eds.), Springer, New York, in press.

10

Experimental Design to Make the Most of Microarray Studies

M. Kathleen Kerr

1. Introduction

Statistics is often thought to concern only the analysis of observational or experimental data. However, experimental design is one of the oldest subfields of statistics. The founder of modern statistics, R. A. Fisher, noted that "statistical procedure and experimental design are only two different aspects of the same whole, and that whole comprises all the logical requirements of the complete process of adding to natural knowledge by experimentation" *(1)*. The design of an experiment affects many things: the analyses that will be possible, the questions that will be answerable, and the quality of the results. While a good design does not guarantee a successful experiment, a suitably bad design guarantees a failed experiment—no results or incorrect results.

Microarrays have been used in some fascinating research, and the technology is tantalizing in the possibilities it presents. In light of this excitement, some perspective is in order. Basically, microarrays are a measurement tool, albeit a high-tech and high-throughput one. With spotted arrays, two differently labeled target DNA samples are simultaneously hybridized to probe DNA immobilized on a microarray. There are many unknown quantities in this process, including the exact sizes and shapes of the probe spots, the density of probe DNA in each spot available for hybridization, and the hybridization efficiency and labeling efficiency of a given sequence. However, regardless of any of these variations, the basic principle is the following: For a given sequence spotted on the array, if one sample contains more of the corresponding transcript, then the signal intensity in the fluor used to label that sample should be proportionately higher than the other fluor. All of the efforts in data normalization are endeavors to return to this principle.

With this perspective, we can see that for a given gene a microarray is really just a comparison between two samples. The fact that the comparison happens for thousands of genes simultaneously introduces some interesting challenges in data analysis. However, it does not have much influence on design considerations.

The following sections outline important aspects of the design of a microarray study. I present general design principles along with guidelines for applying these principles in microarray studies. Everything is discussed in the context of two-color spotted arrays. Most of the principles in **Subheadings 2.**, **3.**, and **5.** also apply directly or indirectly to oligonucleotide arrays.

2. Replication

Replication is, in some sense, the most basic aspect of experimental design. It may be the most widely appreciated—every scientist who performs or requests a sample-size calculation is recognizing the importance of replication. However, there are at least three different kinds of replication with microarrays, with different kinds of importance:

1. Spotting genes multiple times per array.
2. Hybridizing multiple arrays to the same RNA samples.
3. Using multiple individuals of a certain variety or type.

Replication types 1 and 2 are sometimes referred to as experimental or technical replication. These are fundamentally different from type 3. In fact, only type 3 represents replication in the classic statistical sense—random sampling of individuals from a population in order to make inferences about that population.

Replication types 1 and 2 do not address biological variability. Instead, they address the measurement error, or noise in the assay. They may be referred to as *repeated measures (2)* or *subsampling (3,4)* to distinguish them from true replication. Subsampling reduces the uncertainty about gene expression in the particular RNAs in the study. By taking multiple measurements, we hone in on the signal amid the noise. This is particularly useful if the difficulty or cost of sampling additional individuals is much higher than the cost of running additional assays. In addition, spotting genes multiple times per array may serve an important function for the quality control of individual arrays. However, it is generally more advantageous to use true replication, type *(3)*, over subsampling. If subsampling is employed, it will be useful but cannot substitute for true replication to assess biological variability. Many of the most interesting and important scientific questions will require an assessment of biological variability. Experimental variability is important to understand and control but is unrelated to the biology under investigation.

A fundamental property of biological material is that is varies. Genetically identical mice do not respond to treatment exactly the same. Even cell lines vary with culture conditions and the timing of sampling. In pilot studies or cell line studies, an investigator may wish to assume that a single sample is representative of the population. However, the investigator should be aware that he or she is making a critical assumption and, moreover, that the assumption cannot be evaluated with the data. Omitting biological replicates may be reasonable in preliminary studies in which microarrays are used only to identify candidate genes, and extensive follow-up studies are planned. However, in general, microarray studies should assess or account for biological variability with true replication (type 3) in order to produce meaningful scientific results. The difference between assessing and accounting for this variation is discussed in **Subheading 3**.

3. Pooling

Investigators sometimes propose to pool RNA samples from individuals. This strategy has been used successfully in some interesting research *(5)*. Pooling may be wholly or partly motivated by the fact that an insufficient quantity of RNA can be obtained from a single individual to hybridize to an array. If this is the case, then either pooling RNAs or using an RNA amplification procedure are the only courses of action if a microarray study is to be performed. The reliability of RNA amplification protocols is currently being studied.

In other cases, pooling is proposed because biological variation is recognized. Pooling is intended to control or account for this variation. Pooling may be appropriate, but only if the study has limited objectives. Consider a study to compare gene expression changes in mice undergoing a treatment compared with untreated control mice. It would be an interesting scientific finding to identify genes that *on average* are over- or underexpressed in the treated mice. However, describing the difference in two distributions by the difference in their means is a very limited summary. **Figure 1** gives a simple illustration of this point. **Figures 1A** and **1B** each show a pair of normal distributions with the same difference in means. However, there is substantial overlap in the distributions in **Fig. 1A**, in which the distributions have larger variance, and almost no overlap in **Fig. 1B**, in which the variance is much smaller. **Figures 1A** and **1B** illustrate an extremely simple case, when we have two normal distributions with the same variance. In reality, distributions may be skewed or even

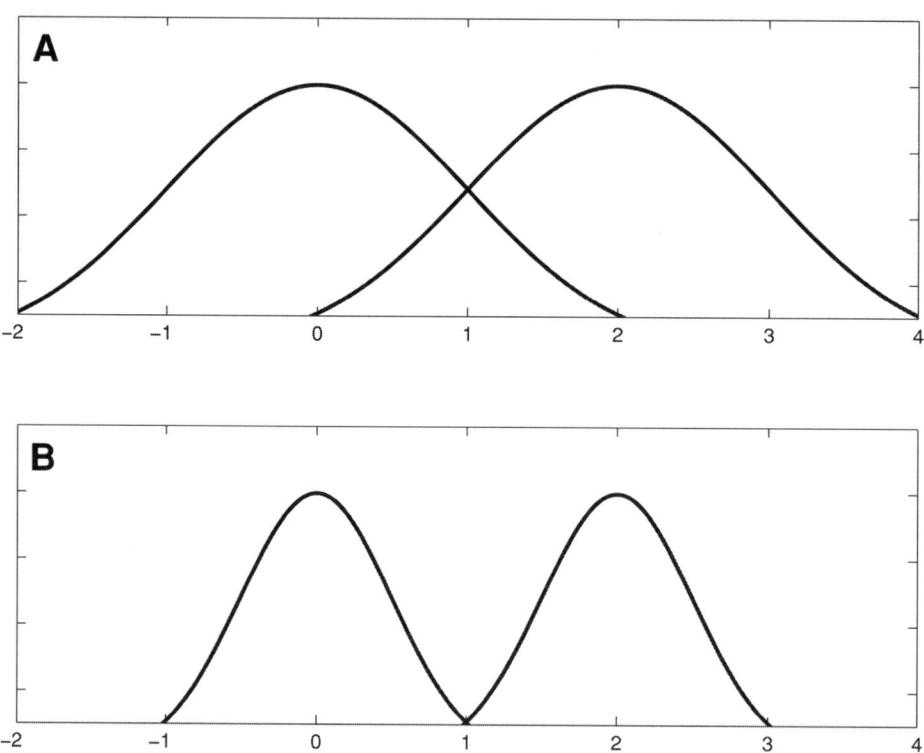

Fig. 1. The pair of distributions in (**A**) have the same difference in means as the pair of distributions in (**B**). However, the variance of the distributions in (**A**) is much larger, so the distributions have substantial overlap.

multimodal. While pooling controls variability, one loses the ability to measure or assess that variability.

If biological variability is assessed, a more careful and complete identification of important genes can be made. Gene A may have a smaller average difference in expression between two groups than gene B, but if the variance is smaller in gene A, it may be considered a more interesting genetic candidate. In some situations, assessing biological variability may be crucial to produce any meaningful results. For example, many microarray studies are directed at finding new diagnostic markers or classification schemes for disease (*6*). The data will be used in a discrimination analysis, or supervised clustering, to examine whether expression data can be used effectively to distinguish diseased samples. Such analyses cannot be executed without information about biological variability. One simply cannot identify the genes that predict disease status based solely on information about mean differences.

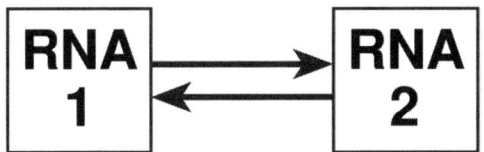

Fig. 2. Graphic representation of a dye-swap experiment. There are two RNA samples and two microarrays; the dye-label assignment is switched between the two arrays (*see* **Table 1**).

Table 1
Dye-Swap Experiment

	Array 1	Array 2
Red channel	RNA 1	RNA 2
Green channel	RNA 2	RNA 1

4. Experimental Layout

A microarray makes a two-sample comparison. A consequence of this is that the experimental layout—how samples are paired onto arrays—is a substantial determinant of design efficiency. Other important concerns are also affected by the layout, such as the ability to discern and pull apart different sources of variation that could otherwise lead to biased results. Because this section is devoted to experimental layout, I first digress to describe the graphic representation of microarray designs.

Microarray design layouts are represented by nodes connected with arrows (7). The nodes represent the RNA samples and the arrows represent the microarrays. Arbitrarily but permanently declare the tail of an arrow to represent one dye and the point of an arrow to represent the other dye. **Figure 2** and **Table 1** show one of the simplest microarray plans, the dyeswap. In a dyeswap there are only two RNA samples and two microarrays. On one array, sample 1 is labeled with one dye and sample 2 is labeled with the other. These dye assignments are then switched on the other array. With larger designs, graphic representations such as **Fig. 2** often communicate layouts more effectively than tables such as **Table 1**, so the graphs are used here.

4.1. Efficiency

As discussed in **Subheading 1.**, microarrays are a comparative measurement tool. The two-dye red/green system addresses many sources of variation that cannot be controlled. The absolute signal intensity from a spot is, by itself, not very informative about gene expression because a high signal can come from high expression but also from a particularly large or dense spot. However, when two differently labeled RNAs are hybridized to the same array, the sample that contains more of a given transcript should produce higher signal in the corresponding spot.

The comparative nature of microarray measurements has led many investigators to routinely use the design depicted in **Fig. 3A**. In the reference design, an additional, arbitrary RNA sample is incorporated into the study as the reference sample. Every sample of interest is compared in a hybridization to this sample. The concept behind the design is simple: every sample of interest can be compared indirectly because each one is compared directly to the reference. For example, if the expression level of a gene is twice as high in sample 1 compared to the reference, and six times as high in sample 2 compared to the reference, then the expression is three times higher in sample 2 compared with sample 1. However, this same simple logic can be applied to more sophisticated designs. As we will see, alternative designs will often have substantial advantages over the reference design. One disadvantage of the reference design is intuitively clear: half of the data are collected on the reference sample, even though this is generally not a sample of interest. A less obvious problem with this design is that it can lack degrees of freedom for estimating error, so one cannot produce error bars on the estimates of relative expression. Without error bars, any kind of rigorous scientific inference is impossible.

One alternative to the reference design is the loop design (7). As depicted in **Fig. 3B**, with this strategy one hybridizes sample 1 with sample 2, sample 2 with sample 3, and so on, until the last sample is assayed with sample 1. The loop design is an example of an even design (*see* **Subheading 4.2.**). If we are interested in all pairwise comparisons of samples in our study (known as A-optimality in statistics), the relative efficiency of the loop to the reference design depends on the number of samples. According to this method of quantifying efficiency, loops are more efficient than reference designs if there are fewer than 10 samples. For example, if there are six samples then a reference design is only 60% as efficient as the corresponding loop design for comparing pairs of samples. This means that, on average, the error bars for relative expression will be $\sqrt{0.6} = 77\%$ as wide if the loop design is used instead of the reference design. This represents an appreciable gain in power for the loop design to detect differential expression without using additional resources. Note that loops automatically incorporate subsampling of the samples of interest—each sample is assayed on two arrays instead of just one in the reference design. However, large loops (more than 10 samples) are less efficient in terms of A-optimality than reference designs. The reason is intuitive: in large loops many pairs of samples are far apart, and, thus, the precision in comparing them is poor.

Experimental Design for Microarrays

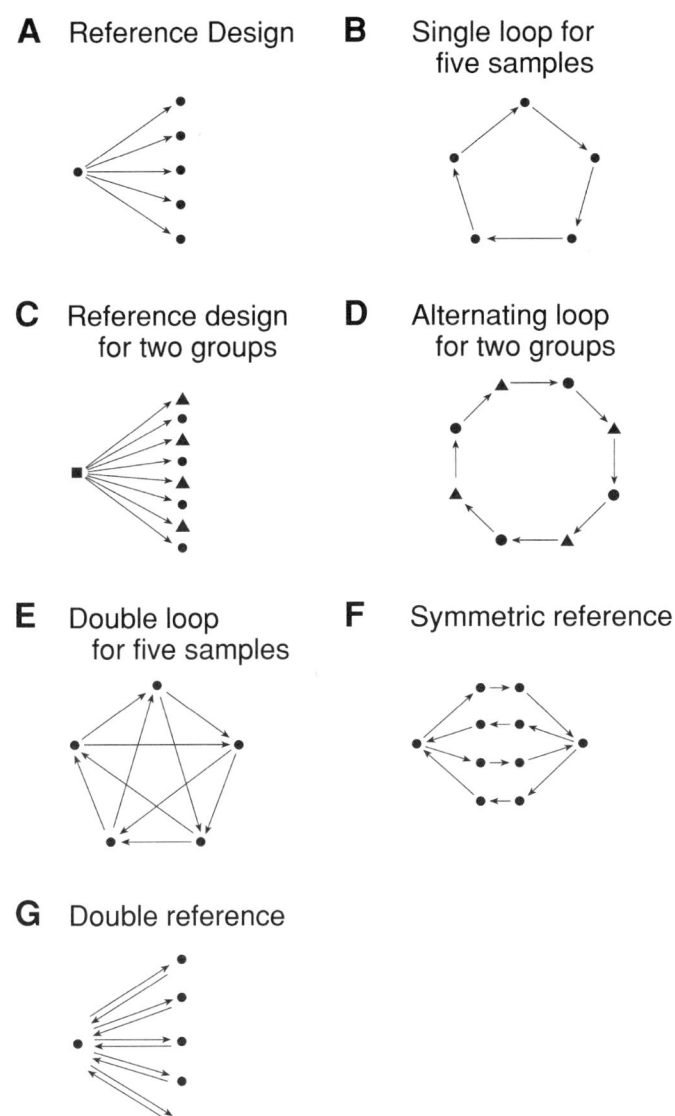

Fig. 3. Assortment of microarray designs. (**A**) Reference design; (**B**) single loop for five samples; (**C**) reference design for two groups; (**D**) alternating loop for two groups; (**E**) double loop for five samples; (**F**) symmetric reference; (**G**) double reference.

But what about replication? The previous evaluation only considers the situation in which we wish to compare all pairs of samples. This criterion is probably not appropriate if we are including biological replicates because we are now interested in comparing sets of samples. Suppose we have N treated mice and *N* control mice and we wish to compare these groups. One option is to use a reference design (**Fig. 3C**). A second option is to use a loop design, alternating mice from each group within the loop (**Fig. 3D**). It turns out that for comparing the two groups, the loop design is always four times more efficient that the reference design, regardless of the number of samples. This means the error bars will be half as wide ($\sqrt{0.25} = 50\%$) for comparing the groups' means if the loop is used instead of the reference design.

The primary goal of this discussion of loop and reference designs is to illustrate the point that the design layout is a major determinant of estimation precision and therefore the statistical power of a study. However, there are many designs other than loop designs and reference designs and many cases in which neither design is a good choice. Simply stated, there is no universal microarray design *(8)*. Kerr and Churchill *(7)* performed an exhaustive search of A-optimal designs for 10 or fewer samples and different numbers of arrays. They described families of good designs, some of which could be described as double loops (**Fig. 3E**) or symmetric reference designs (**Fig. 3F**). Although there is no simple formula for constructing a good design or calculating design efficiency, an effective guideline is to construct the experimental layout so that samples to be compared are closely linked in the design.

4.2. Dye Bias

Some researchers have observed gene-specific dye biases in their microarray data. This is sometimes described as observing genes for which "the ratios do not flip." This bias is easily seen in simple dye-swap experiments, in which two samples are hybridized onto two arrays, and the dye labels are switched in the two hybridizations (**Fig. 1, Table 1**). After normalizing for other sources of variation, including overall differences in the dyes, one expects the ratio green/red from array 1 to be about the same as the ratio red/green from array 2 for any particular gene. For a subset of genes, instead one finds that green/red from array 1 is about the same as green/red from array 2. When this bias is observed, it is typically <1% of genes, which means up to 100 genes if 10,000 are spotted.

The source of this dye bias is currently under investigation, though it is generally suspected to be an artifact of some dye-labeling protocols. Fortunately, it is fairly straightforward for investigators to protect themselves from being misled by this bias until the issue is resolved. An effective way to account for dye bias is to dye swap every assay. An example of such a design is the double reference design in **Fig. 3G**. While dye swapping every assay adequately addresses the problem, it is not necessary; it is sufficient to use a design that is even. An *even design (7)* is one in which every sample is labeled with both dyes, and each differently labeled subsample is used equally often in the experimental layout. The designs in **Fig. 2**, **3B**, and **D–G** are all even designs. In statistical jargon, even designs are designs in which gene expression effects and dye effects are *orthogonal*. If an even design is used, dye biases will not affect estimates of relative gene expression because the dyes are balanced across samples. At the other extreme, a design such as the reference design in **Fig. 3C** has expression effects and dye effects completely *confounded*. In a confounded design such as the reference design, dye biases cannot be corrected. In fact, these biases cannot even be detected. This means that an investigator cannot know whether his or her results are biased. Even designs protect against this bias, without the potentially large cost of performing a dye swap for every array.

4.3. Robustness

Robustness is an additional important design consideration. Generally speaking, robustness in design refers to the relative efficiency of the effective design if there are missing data due to failed arrays or bad spots on some arrays in the intended design. The loop design (**Fig. 3B, D**) is not a robust design. If a loop design is planned, but data are not obtained from some array, the resulting effective design has greatly reduced efficiency for all comparisons. Worse, if data are lost on multiple arrays, the design may become disconnected, and some samples then cannot be compared at all. By contrast, the reference design (**Fig. 3A**) is robust. A reference design remains a reference design if an array is lost, and the lost data affect only comparisons with the sample on that array.

Loops are not a robust design, which is a compelling argument against them for labs that commonly have failed or partially failed assays that cannot be replaced. However, variations on loops are very robust designs. Compare the double-loop and double-reference designs in **Fig. 3E** and **3G**. These two designs each use $2n$ arrays to study n samples. For five samples, the relative efficiency of these designs is 40% in favor of the double loop. In addition, the double loop is more robust, because there are many more connections between any two samples, so many data remain to compare any pair of samples if some data are lost. By contrast, losing a single array in the double-reference design means half the data on a particular sample are lost.

4.5. Other Considerations

In addition to the considerations of efficiency and dye bias, there may be other important qualities in an experimental layout. These considerations include simplicity, extendibility, and useful subdesigns. Simplicity may be important for a large study in which many technicians will be involved in performing the assays. If a study is somewhat open ended, a design that is easily extendable may be preferred. Extendability means that if a new sample is acquired late in the study, it can be added to the design in a sensible way. Double reference (**Fig. 3G**) and symmetric reference (**Fig. 3F**) are natural to extend to include additional samples. Finally, it may be desirable for a design to contain useful subdesigns. If a question of interest applies to only a subset of samples, then it may be useful to have a subdesign of arrays that is efficient for addressing that question. This will allow the question to be studied by analyzing only a subset of the data.

5. Randomization

One aspect of design that has not been mentioned up to this point is randomization. In general, randomization is an important component of sound experimentation. In a microarray study, many opportunities for randomization may appear before any array hybridizations. For example, if the effects of two treatments on a mouse model are to be compared, then mice should be randomly assigned into the two treatment groups. For any set of microarray assays, arrays should be assigned randomly to positions in the chosen design from the batch of arrays to be used. This will control any systematic bias due to the order in which the arrays were printed or manufactured. Ideally, randomization would be employed at other stages of the process, but, unfortunately, this is often not possible. For example, it would be preferable if the location of the genes could be randomized from array to array so that any systematic spatial variation would not bias the results. Of course, this is not practical for the robots that print the arrays. However, a general principle of scientific experimentation is to randomize whenever possible to protect against unanticipated biases.

6. Conclusions

In a study employing spotted microarrays, these are usually the two most important design questions: Which RNA samples will be collected for the assays? and Which experimental layout will be used? The scientific question of interest should drive the choices in each case.

Replication is the important component of selecting the RNA samples. Without true replicates (type 3 in **Subheading 2.**), biological variability cannot be assessed and the desired inferences cannot be made. Certain overarching goals of the study, such as finding a classification scheme, may not be possible without biological replication. Subsampling, or technical replication (types 1 and 2 in **Subheading 2.**), can be useful but can never substitute for true replication.

A spotted microarray is effectively a comparison between two samples. Because of this, the way that samples are paired onto arrays has a huge impact on how effectively one can make all of the comparisons of interest at the end of the study. The general rule of thumb is that samples to be compared should be close in the design. The experimental layout also determines whether the effects of interest (i.e., gene expression effects) are confounded with other sources of variation. In particular, if aberrant dye effects are a concern, then it is important to use an even design so that these will not bias the results. Otherwise, these effects may be confounded with gene expression changes, and there will be no way to tell whether end results are biased or not.

Any finite body of data contains a finite amount of information. The information content of a data set is determined by the design of the experiment that produced it, regardless of the particular data values that appear. Once a data set is collected, its information content cannot be increased by any amount of ingenuity expended by a data analyst *(1)*. Good design is crucial to all scientific experimentation, and microarrays are no exception.

References

1. Fisher, R. A. (1971) *The Design of Experiments*, 8th ed., Hafner, New York.
2. Kerr, M. K. and Churchill, G. A. (2001) Statistical design and the analysis of gene expression microarray data. *Genet. Res.* **77,** 123–128.
3. Snedecor, G. W. (1989) *Statistical Methods*, 8th ed., Iowa State University Press, Ames, IA.
4. Kerr, M. K., Afshari, C. A., Bennett, L., Bushel, P., Martinez, J., Walker, N. J., and Churchill, G. A. (2002) Statistical analysis of a gene expression microarray experiment with replication. *Statistica Sinica* **12,** 203–217.
5. Jin, W., Riley, R. M., Wolfinger, R. D., White, K. P., Passador-Gurgel, G., and Gibson, G. (2001) The contributions of sex, genotype and age to transcriptional variance in Drosophila melanogaster. *Nat. Genet.* **29,** 389–395.
6. Dudoit, S., Fridlyand, J., and Speed, T. P. Comparison of discrimination methods for the classification of tumors using gene expression data. *J. Am. Stat. Assoc.* **97,** 77–87.
7. Kerr, M. K. and Churchill, G. A. (2001) Experimental design for gene expression microarrays. *Biostatistics* **2,** 183–201.
8. Churchill, G. A. and Oliver, B. (2001) Sex, flies, and microarrays. *Nat. Genet.* **29,** 355–356.

11

Statistical Methods for Identifying Differentially Expressed Genes in DNA Microarrays

John D. Storey and Robert Tibshirani

1. Introduction

In this chapter we discuss the problem of identifying differentially expressed genes from a set of microarray experiments. Statistically speaking, this task falls under the heading of "multiple hypothesis testing." In other words, we must perform hypothesis tests on all genes simultaneously to determine whether each one is differentially expressed. Recall that in statistical hypothesis testing, we test a null hypothesis vs an alternative hypothesis. In this example, the null hypothesis is that there is no change in expression levels between experimental conditions. The alternative hypothesis is that there is some change. We reject the null hypothesis if there is enough evidence in favor of the alternative. This amounts to rejecting the null hypothesis if its corresponding statistic falls into some predetermined rejection region. Hypothesis testing is also concerned with measuring the probability of rejecting the null hypothesis when it is really true (called a *false positive*), and the probability of rejecting the null hypothesis when the alternative hypothesis is really true (called *power*).

There are three important steps one must take in testing for differential gene expression. The first is that a statistic must be formed for each gene. The choice of this statistic is important in that one wishes to ensure that no relevant information is lost with respect to the test of interest, yet that all measurements on the gene are condensed into one number. The second step is choosing the rejection regions. This involves comparing the original statistics to null versions of the statistics. Asymmetric rejection regions are most appropriate because we do not know beforehand what proportion of differentially expressed genes is in the positive or negative direction. If either of these two steps is not performed well, a loss in statistical power can result.

The third main challenge is to assess or control the number of false positives. In multiple hypothesis testing, the familywise error rate (FWER) has traditionally been the main quantity of interest. The FWER is the probability of one or more false positives occurring among all significant hypotheses. However, in microarray studies, the number of tests (genes) is large, and the nature of the analysis is exploratory; that is, we want to identify many differentially expressed genes, without too many false positives resulting. Hence, we expect many more than one false positive, but we do not want too many in proportion to true positives. Therefore, the FWER is not a relevant quantity in this setting. Instead, we focus here on the *false discovery rate* (FDR), defined to be the expected proportion of false positive genes among all those called significant.

In the remainder of this chapter, we discuss some simple methods for achieving these three steps in testing for differential gene expression. For simplicity, we limit our discussion to the case in which there are two experimental conditions and each sample is independent.

2. Forming the Test Statistics

Suppose we have m genes measured on n arrays, under two different experimental conditions. Let $\bar{x}_{i1}$ and $\bar{x}_{i2}$ be the average gene expression for gene i under conditions 1 and 2, and let s_i be the pooled standard deviation for gene i:

$$s_i = \left[\left(\frac{1}{n_1} + \frac{1}{n_2} \right) \cdot \sum_1 (x_{ij} - \bar{x}_{i1})^2 + \sum_2 (x_{ij} - \bar{x}_{i2})^2 \right]^{1/2}$$

Here, n_k is the number of arrays in condition k, and each summation is taken over its respective group. Then a reasonable test statistic for assessing differential gene expression is the standard (unpaired) t statistic:

$$d_i = \frac{\bar{x}_{i2} - \bar{x}_{i1}}{s_i}$$

Another statistic that could be formed is the rank-sum statistic. Let r_{ij} for $j = 1, \ldots, n$ be the rank of the jth expression value within gene i. Then the rank-sum statistic for gene i is:

$$r_i = \sum_1 r_{ij}$$

in which the summation is taken over the genes in group 1. An extreme r_i value in either direction would indicate a difference in gene expression. In most cases, the statistic d_i is better than r_i because much information is lost in r_i.

Table 1
Possible Outcomes from m Hypothesis Tests

	Accept	Reject	Total
Null true	U	V	m_0
Alternative true	T	S	m_1
	W	R	m

Sometimes one wishes to test for a difference among three or more conditions, or for a trend in time-course data. Statistics that can be used for these and other situations are described in **ref. 1**.

3. Thresholding and the FDR

Using the statistic d_i, we can simply compute its value for each gene, choose a threshold t, and then declare significant all genes satisfying $|d_i| > t$.

What are the operating characteristics of this procedure? For an individual gene, we could estimate its significance from a normal table; for example, if $t = 1.5$, the (two-tailed) significance level (or p value) would be 13.4%. However, it is more robust to use permutations to assess the significance: we randomly scramble the set of n treatment labels, and recompute the d_i value for gene i. We do this, say, 100 times, and compute the proportion among the 100 that exceeds the original d_i in absolute value. If, say, 10% of the values exceed the original value, the significance level would be 10%.

Then, V is the number of false positives (type I errors), while R is the total number of hypotheses rejected. The FWER is defined to be $\Pr(V \geq 1)$, and the FDR is the expected value of V/R.

This works well for one gene, but we wish to test many genes all at once. Consider **Table 1**, which displays the various outcomes when testing m genes. The well-known Bonferonni method provides a way of choosing t to control the FWER. For example, to control it at level 0.10, we choose t so that $\Pr(|d_i| > t) = 0.10/m$. Unfortunately, since m is very large here ($m = 3000$), the resulting threshold t is so large that few or no genes will be called significant. Other methods have been proposed for controlling the FWER (e.g., *see* **ref. 2**), but these suffer from the same problem. In the next example, we describe the FDR approach.

4. An Example

We obtained expression data from Stanford collaborators on 3000 genes, comparing normal untreated samples with a set of treated samples. (Since the data are unpublished, we must omit the biological details.) There are

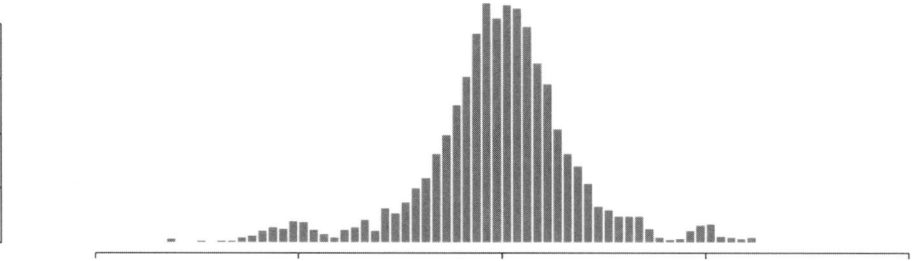

Fig. 1. Histogram of 3000 t statistics, from DNA microarray example.

15 samples in group 1 (normal) and 13 in group 2 (treated). **Figure 1** is a histogram of the 3000 two-sample t statistics from the genes. They range from –4.54 to 3.72. Suppose we decide to reject all genes whose t statistic is >1.5 in absolute value; there are 252 such genes. What is the FDR among these 252 genes?

To assess this, we do a random permutation of the sample labels

$$(1,1,1,1,1,1,1,1,1,1,1,1,1,1,1,2,2,2,2,2,2,2,2,2,2,2,2,2)$$

recompute the t statistics, and count how many exceed ±2. Doing this for 100 permutations, we find the average number is 64.4. Thus, a simple estimate of the FDR is 64.4/252=25.5%.

Now it turns out that this simple estimate tends to be biased upward. The reason is clear from **Table 1**. The permutations make all m genes null, but in our data only a proportion $\pi_0 = m_0/m$ are null. Hence, to improve our estimate of the FDR, we multiply it by an estimate of π_0. To obtain the latter, we look at small values of the t statistic (in absolute value), where null statistics are much more abundant than alternatives. Looking for an example at t statistics below 0.15 in absolute value, we find that 668 of the observed t statistics fall in that range, while on the average 750 of the t statistics from the permutations fall in that range. Hence, our estimate of π_0 is $\pi_0 = 668/750 \approx 0.890$. Finally, our revised estimate of the FDR is $0.890 \cdot 25.5 = 22.7\%$.

The FDR is appropriate for detecting differential gene expression, as mentioned previously. For example, if we estimate the FDR to be 5% among a list of 100 significant genes, then we roughly expect that 5 of those 100 are false positives. This kind of information is very useful for choosing genes to examine more carefully, and it is much more statistically rigorous than taking the top k genes for some arbitrarily chosen k. The FDR was first introduced by Benjamini and Hochberg (3); in **refs. 4** and **5** we have developed more accurate ways of estimating this quantity.

Identifying Differentially Expressed Genes

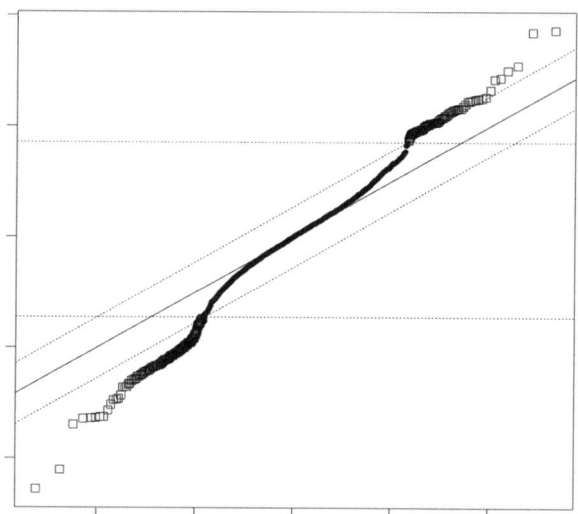

Fig. 2. SAM plot for microarray example. We draw a band of two parallel (broken) lines a distance $\Delta = 0.531$ from the 45° (solid) line. Moving up and to the right, we find the first time that the points go outside the band. All genes to the right of that point are called significant (even if they fall inside the band). We do the same thing in the bottom left corner. The upper and lower excursion values imply upper and lower cut points—here 1.71 and −1.45. These cut points are then used in estimating the FDR from permuted data sets.

5. Improving Cut Points

The significance analysis of microarrays (SAM) method and software *(1)* forms cut points based on the data. Rather than using a standard rule of the form $|d_i| > t$ to call genes significant (i.e., having symmetric cut points $\pm t$) SAM derives cut points t_1 and t_2 and uses the rejection rule $d_i < t_1$ or $d_i > t_2$. This can lead to a more powerful test in situations in which more genes are overexpressed than underexpressed, or vice versa. Given its cut points, the SAM method estimates the FDR as described next. The derivation of the cut points is also described next as well as illustrated in **Fig. 2**.

5.1. The SAM Procedure

1. Compute the ordered statistics $d_{(1)} \leq d_{(2)} \ldots \leq d_{(m)}$.
2. Take B sets of permutations of the group labels. For each permutation b, compute statistics d_i^{*b} and corresponding order statistics $d_{(1)}^{*b} \leq d_{(2)}^{*b} \leq \cdots d_{(m)}^{*b}$. From the set of B permutations, estimate the expected order statistics by $\bar{d}_{(i)} = (1/B)\Sigma_{b=1}^{B} d_{(1)}^{*b}$ for $i = 1, 2, \ldots, m$.

3. Plot the $d_{(i)}$ values vs the $\bar{d}_{(i)}$. For a fixed threshold Δ, starting at the origin, and moving up to the right, find the first $i = i_1$ such that $d_{(i)} - \bar{d}_{(i)} > \Delta$. All genes past i_1 are called *significant positive*. Similarly, start at the origin, move down to the left, and find first $i = i_2$ such that $\bar{d}_{(i)} - d_{(i)} > \Delta$. All genes past i_2 are called *significant negative*. For each Δ define the upper cut point t_2 as the smallest d_i among the significant positive genes and similarly define the lower cut point t_1.

In **Fig. 2** we used $\Delta = 0.531$, to yield the same number of genes (252) that we obtained using the cutoffs ± 1.5. As **Fig. 2** shows, the resulting cut points from SAM are asymmetric; they are -1.45 and 1.71. As a result, the number of genes called significant at the negative and positive ends is quite unequal—they are 167 and 85 (and they are 157 and 95 for the cut points ± 1.5).

The actual SAM procedure adds (the same) constant s_0 to the denominator of each statistic d_i. The constant is chosen as the percentile of the s_i values that makes the coefficient of variation of d_i approximately constant as a function of s_i. Its effect is to dampen large values of d_i that arise from genes whose expression is near zero. Here, we set s_0 equal to the median value of the s_i, 17.7. For consistency in this example, we used this value in the denominator of all t statistics for all methods.

In the example in this section, SAM chose cut points $t_1 = -1.45$ and $t_2 = 1.71$ to produce 252 significant genes. The algorithm in the following section explains how to estimate the FDR for the gene list produced from the SAM cut points. It is the same procedure as described for the symmetric cut points. For concreteness we outline it in detail.

5.2. Estimation of FDR for SAM Procedure

1. Calculate the statistics $d_1, \ldots, d_m$ as outlined above.
2. Choose a set of cut points t_1 and t_2 based on some method, such as SAM.
3. Let t_1^0 and t_2^0 be two additionally chosen cut points so that the statistics between t_1^0 and t_2^0 are mostly null. (This choice is fully automated in the SAM software based on **ref. 5**.)
4. Using B null versions of statistics $d_1^{*b}, d_2^{*b}, \ldots, d_m^{*b}$ for $b = 1, \ldots, B$ estimate π_0 by

$$\hat{\pi}_0 = \frac{\text{proportion of } d_i \text{ such that } t_1^0 \leq d_i \leq t_2^0}{\text{proportion of } d_i^{*b} \text{ such that } t_1^0 \leq d_i^{*b} \leq t_2^0 \text{ over } b = 1, \ldots, B}$$

5. Estimate the FDR by

$$\widehat{FDR} = \hat{\pi}_0 \cdot \frac{\text{proportion of } d_i^{*b} \text{ such that } d_i^{*b} < t_1 \text{ or } d_i^{*b} > t_2 \text{ over } b = 1, \ldots, B}{\text{proportion of } d_i \text{ such that } d_i < t_1 \text{ or } d_i > t_2}$$

For our data, there were an average of 54.76 false positives in the 100 permuted data sets, so using the estimate $\hat{\pi}_0 = 0.89$ from earlier, our estimate of FDR is $0.89 \cdot 54.76/252 = 19.3\%$. This is a little lower than the estimate of

22.7% for the symmetric rule. This often happens—by using the asymmetric cut points from SAM, we obtain lower FDRs than are obtained from symmetric ones.

Note that the value $\Delta = 0.531$ yielding 252 genes was chosen somewhat arbitrarily in the previous example. In practice, the investigator will want to vary Δ and examine the number of significant genes and the FDR (*see* **Subheading 6.**). In some settings, a short list of significant genes is desired, such as when Northern blot tests are to be carried out on the individual genes. In other more exploratory settings, a longer gene list would be preferred.

It is shown in **refs. 4** and **5** that the preceding estimate of FDR provides strong control of the FDR in that $E[\widehat{FDR}] \geq FDR$ for any m or π_0. That is, we expect that our estimate is conservative on average. Tusher et al. *(1)* use the same estimate of the FDR, except without the $\hat{\pi}_0$ term. Other methods such as in **refs. 6** and **7** provide methods for controlling the FDR when dependence exists, but it is shown in **ref. 5** that our methodology is more powerful. (Note that there is dependence between the tests because the genes can be dependent.) In addition, Dudoit et al. *(8)* suggest using the methodology of Westfall et al. *(2)* to control the FWER when testing for differential gene expression. To reject the 252 genes that we did, the FWER would have to be controlled at a level of 99.8% using the methodology in **ref. 2**. This clearly shows that the FWER is an undesirable quantity to control.

6. The *q* Value

A natural question to ask is how does one pick the rejection region (Δ value)? In single hypothesis testing, one can assign a *p* value to a statistic, which is usually defined to be the probability of a statistic more extreme than the one observed given that the null hypothesis is true. The more technical definition of the *p* value for an observed statistic is the minimum type I error rate that can be attained over all rejection regions containing that observed statistic. In **ref. 9**, a natural false discovery rate analog of the *p* value is introduced, which is called the *q* value. The *q* value of a statistic is the minimum positive FDR (pFDR) that can be attained over all rejection regions containing that observed statistic. The pFDR is a modified version of the FDR—in short, it conditions on the event that $R > 0$ rather than setting $V/R = 0$ when $R = 0$. (We argue *[9]* that the pFDR is a more appropriate error rate than the FDR.) Therefore, in testing for differential gene expression, we suggest calculating the *q* value for each gene. It gives a measure of the strength of evidence for differential gene expression in terms of the pFDR. This is an individual measure for each gene that simultaneously takes into account the multiple comparisons. Note that by using the *q* value, it is not necessary to pick the rejection region or the desired error rate beforehand. The *q* value is included in the SAM software.

7. Discussion

Several other methods have been suggested for detecting differential gene expression. Newton et al. *(10)* developed a Bayesian method, although the multiple hypothesis testing was not taken into account. In fact, only posterior probabilities are reported according to this methodology. However, Newton et al. *(10)* do suggest taking into account both fold change and the variance of the fold change (equivalent to a t statistic). Efron et al. *(11)* suggest an empirical Bayes approach. Posterior probabilities are calculated as well, and no distributional assumptions are made. In both of these methods, it is important to realize that a posterior probability of differential gene expression given the value of the t statistic for only the gene at hand is not sufficient for looking at several genes at once. In **ref. 12**, Efron et al. go further, relating empirical Bayes probabilities to global and local versions of the FDR. In **ref. 8**, symmetric rejection regions and the FWER as the multiple hypothesis testing measure are used. They use the adjusted p value methodology of **ref. 2**. We have already discussed why symmetric rejection regions and the FWER result in a significant loss of power to detect differential gene expression. And we see no need to compute p values and adjust them. In this chapter, we have worked directly with test statistics and FDRs.

A strength of the approach we have presented is that the rejection regions are chosen to accommodate the proportion and degree to which the affected genes are over- or underexpressed. In addition, the multiple comparisons are taken into account via a multiple hypothesis testing error rate that is appropriate for finding many significant genes, while limiting the proportion of false positives among those called significant. Of course, other methods may exist for choosing rejection regions or for assessing the false positives that provide a more powerful, but just as efficient, detection of differential gene expression. This is an active area of research, and some additional, interesting statistical methodology is bound to emerge.

The methodology presented in this chapter is implemented in the SAM software and is available for downloading at www-stat.stanford.edu/~tibs/SAM/.

References

1. Tusher, V., Tibshirani, R., and Chu, C. (2001) Significance analysis of microarrays applied to transcriptional responses to ionizing radiation. *Proc. Natl. Acad. Sci. USA* **98,** 5116–5121.
2. Westfall, P. H. and Young, S. S. (1993) *Resampling-Based Multiple Testing: Examples and Methods for p-Value Adjustment*, Wiley, New York.
3. Benjamini, Y. and Hochberg, Y. (1985) Controlling the false discovery rate: a practical and powerful approach to multiple testing. *J. Roy. Stat. Soc. B* **85,** 289–300.

4. Storey, J. D. A direct approach to false discovery rates, submitted. Available at www-stat.stanford.edu/~jstorey/.
5. Storey, J. D. and Tibshirani, R. Estimating false discovery rates under dependence, with applications to DNA microarrays, submitted. Available at www-stat.stanford.edu/~jstorey/.
6. Yekutieli, D. and Benjamini, Y. (1999) Resampling-based false discovery rate controlling multiple test procedures for corelated test statistics. *J. Stat. Plan. Infer.* **82,** 171–196.
7. Benjamini, Y. and Yekutieli, D. The control of the false discovery rate in multiple testing under dependency, in press.
8. Dudoit, S., Yang, Y., Callow, M., and Speed, T. Statistical methods for identifying differentially expressed genes in replicated cdna microarray experiments. Available at www.stat.berkeley.edu/users/sandrine.
9. Storey, J. D. The positive false discovery rate: a Bayesian interpretation and the q-value, submitted. Available at www-stat.stanford.edu/~jstorey/.
10. Newton, M., Kendziorski, C., Richmond, C., Blatter, F., and Tsui, K. (2001) On differential variability of expression ratios: improving statistical inference about gene expression changes from microarray data. *J. Compu. Biol.* **8,** 37–52.
11. Efron, B., Tibshirani, R., Storey, J. D., and Tusher, V. Empirical Bayes analysis of a microarray experiment. *J. Am. Stat. Assoc.*, in press.
12. Efron, B., Storey, J., and Tibshirani, R. Microarrays, empirical Bayes methods, and false discovery rates, submitted.

12

Detecting Stable Clusters Using Principal Component Analysis

Asa Ben-Hur and Isabelle Guyon

1. Introduction

Clustering is one of the most commonly used tools in the analysis of gene expression data *(1,2)*. The usage in grouping genes is based on the premise that coexpression is a result of coregulation. It is often used as a preliminary step in extracting gene networks and inference of gene function *(3,4)*. Clustering of experiments can be used to discover novel phenotypic aspects of cells and tissues *(3,5,6)*, including sensitivity to drugs *(7)*, and can also detect artifacts of experimental conditions *(8)*. Clustering and its applications in biology are presented in greater detail in Chapter 13 (see also **ref. 9**). While we focus on gene expression data in this chapter, the methodology presented here is applicable for other types of data as well.

Clustering is a form of unsupervised learning; that is, no information on the class variable is assumed, and the objective is to find the "natural" groups in the data. However, most clustering algorithms generate a clustering even if the data have no inherent cluster structure, so external validation tools are required. Given a set of partitions of the data into an increasing number of clusters (e.g., by a hierarchical clustering algorithm, or k-means), such a validation tool will tell the user the number of clusters in the data (if any). Many methods have been proposed in the literature to address this problem *(10–15)*. Recent studies have shown the advantages of sampling-based methods *(12,14)*. These methods are based on the idea that when a partition has captured the structure in the data, this partition should be stable with respect to perturbation of the data. Bittner et al. *(16)* used a similar approach to validate clusters representing gene expression of melanoma patients.

From: *Methods in Molecular Biology: vol. 224: Functional Genomics: Methods and Protocols*
Edited by: M. J. Brownstein and A. Khodursky © Humana Press Inc., Totowa, NJ

The emergence of cluster structure depends on several choices: data representation and normalization, choice of a similarity measure and clustering algorithm. In this chapter, we extend the stability-based validation of cluster structure and propose stability as a figure of merit that is useful for comparing clustering solutions, thus helping one to make these choices. We use this framework to demonstrate the ability of principal component analysis (PCA) to extract features relevant to the cluster structure. We use stability as a tool for simultaneously choosing the number of principal components (PCs) and the number of clusters; we compare the performance of different similarity measures and normalization schemes. The approach is demonstrated through a case study of yeast gene expression data from Eisen et al. *(1)*. For yeast, a functional classification of a large number of genes is known, and we use this classification for validating the results produced by clustering. A method for comparing clustering solutions specifically applicable to gene expression data was introduced in **ref. *17***. However, it cannot be used to choose the number of clusters and is not directly applicable in choosing the number of PCs.

The results of clustering are easily corrupted by the addition of noise: even a few noise variables can corrupt a clear cluster structure *(18)*. Several factors can hide a cluster structure in the context of gene expression data or other types of data: the cluster structure may be apparent in only a subset of the experiments or genes, or the data themselves may be noisy. Thus, clustering can benefit from a preprocessing step of feature/variable selection or from a filtering or denoising step. In the gene expression case study presented in this chapter, we find that using a few leading PCs enhances cluster structure. For a recent article that discusses PCA in the context of clustering gene expression *see* **ref. *19***.

PCA constructs a set of uncorrelated directions that are ordered by their variance *(20,21)*. In many cases, directions with the most variance are the most relevant to the clustering. Our results indicate that removing features with low variance acts as a filter that results in a distance metric that provides a more robust clustering. PCA is the basis for several variable selection techniques; variables that have a large component in low variance directions are discarded *(22,23)*. This is also the basis of the "gene shaving" method *(24)* that builds a set of variables by iteratively discarding variables that are least correlated with the leading PCs. In the case of temporal gene expression data, the PCs were found to have a biological meaning, with the first components having a common variability *(25–27)*. PCA is also useful as a visualization tool; it can provide a low-dimensional summary of the data *(28)*, help detect outliers, and perform quality control *(20)*.

2. Principal Components

We begin by introducing some notation. Our object of study is an n by d gene expression matrix, X, giving the expression of n genes in d experiments. The gene expression matrix has a dual nature: one can cluster either genes or experiments; to express this duality we can refer to X as

$$X = \begin{pmatrix} - & g_1 & - \\ & \vdots & \\ - & g_n & - \end{pmatrix} \quad (1)$$

in which $g_i = (x_{i1}, \ldots, x_{id})$ are the expression levels of gene i across all experiments, or as

$$X = \begin{pmatrix} | & & | \\ e_1 & \cdots & e_d \\ | & & | \end{pmatrix} \quad (2)$$

in which $e_j = (x_{1j}, \ldots, x_{nj})'$ are the expression levels in experiment j across all genes. In this chapter we cluster genes, i.e., the n patterns g_i. When clustering experiments, substitute e and g in what follows. We make a distinction between a *variable*, which is each one of the d variables that make up g_i, and a *feature*, which denotes a combination of variables.

The PCs are q orthogonal directions that can be defined in several equivalent ways *(20,21)*. They can be defined as the q leading eigenvectors of the covariance matrix of X. The eigenvalue associated with each vector is the variance in that direction. Thus, PCA finds a set of directions that explains the most variance. For Gaussian data, the PCs are the axes of any equiprobability ellipsoid. A low-dimensional representation of a data set is obtained by projecting the data on a small number of PCs. Readers interested in theoretical or algorithmic aspects of PCA should refer to textbooks devoted to the subject *(20,21)*.

The PCs can be defined as the q leading eigenvectors of the experiment–experiment covariance matrix:

$$Cov(X)_{ij} = \frac{1}{n}(e_i - <e_i>)'(e_j - <e_j>), \quad i, j = 1, \ldots, d \quad (3)$$

in which $<e_j> = 1/n \sum_{i=1}^{n} x_{ij}(1, \ldots, 1)$ is a d-dimensional vector with the mean expression value for experiment j. Alternatively, the PCs can be defined through the correlation matrix:

$$Cor(X)_{ij} = \frac{1}{n}\frac{(e_i - <e_i>)'}{\sigma(e_i)}\frac{(e_j - <e_j>)}{\sigma(e_j)}, \quad i, j = 1, \ldots, d \quad (4)$$

in which $\sigma(e_i)$ is the vector of estimated standard deviation in experiment i. Equivalently, one can consider the PCs as the eigenvectors of the matrix XX' when applying first a normalization stage of *centering*:

$$e_i \rightarrow e_i - <e_i> \quad (5)$$

or *standardization*:

$$e_i \rightarrow (e_i - <e_i>)/\sigma(e_i) \quad (6)$$

The first corresponds to PCA relative to the covariance matrix, and the second to PCA relative to the correlation matrix. To distinguish between the two, we denote them by centered PCA and standardized PCA, respectively. One can also consider PCs relative to the second moment matrix, i.e., without any normalization. In this case, the first PC often represents the mean of the data, and the larger the mean, the larger this component relative to the others.

We note that for the case of two dimensional data the correlation matrix is of the form $(1, a; a, 1)$, which has *fixed* eigenvectors $(x, -x)$ and (x, x) ($x = \sqrt{2}/2$, for normalization), regardless of the value of a. Standardization can be viewed as putting constraints on the structure of the covariance matrix that in two dimensions fixes the PCs. In high-dimensional data this is not an issue, but a low-dimensional comparison of centered PCA and standardized PCA would be misleading. Standardization is often performed on data that contain incommensurate variables (i.e., variables that measure different quantities) and are incomparable unless they are made dimensionless (e.g., by standardization). In the case of commensurate variables, it has been observed that standardization can reduce the quality of a clustering *(29)*. In microarray data, all experiments measure the same quantity, namely mRNA concentration, but still normalization across experiments might be necessary.

The basic assumption in using PCA as a preprocessing before clustering is that directions of large variance are the result of structure in those directions. We begin with a toy example that illustrates this, and later we show results for gene expression data for which this applies as well. Consider the data plotted in **Fig. 1** (see caption for details on its construction). There is clear cluster structure in the first and second variables. Centered PCA was applied. The first PC captures the structure that is present in the first two variables. **Figure 2** shows that this component is essentially a 45° rotation of the first two variables.

3. Clustering and Hierarchical Clustering

All clustering algorithms use a *similarity* or *dissimilarity* matrix and group patterns that are similar to each other. A similarity matrix gives a high score to "similar" patterns, with common examples being the Euclidean dot product or Pearson correlation. A dissimilarity matrix is a matrix whose entries reflect a "distance" between pairs of patterns; that is, close patterns have a low dissimilarity. The input to the clustering algorithm is either the similarity/dissimilarity matrix or the data patterns themselves, and the elements of the matrix are computed as needed.

Detecting Stable Clusters Using PCA

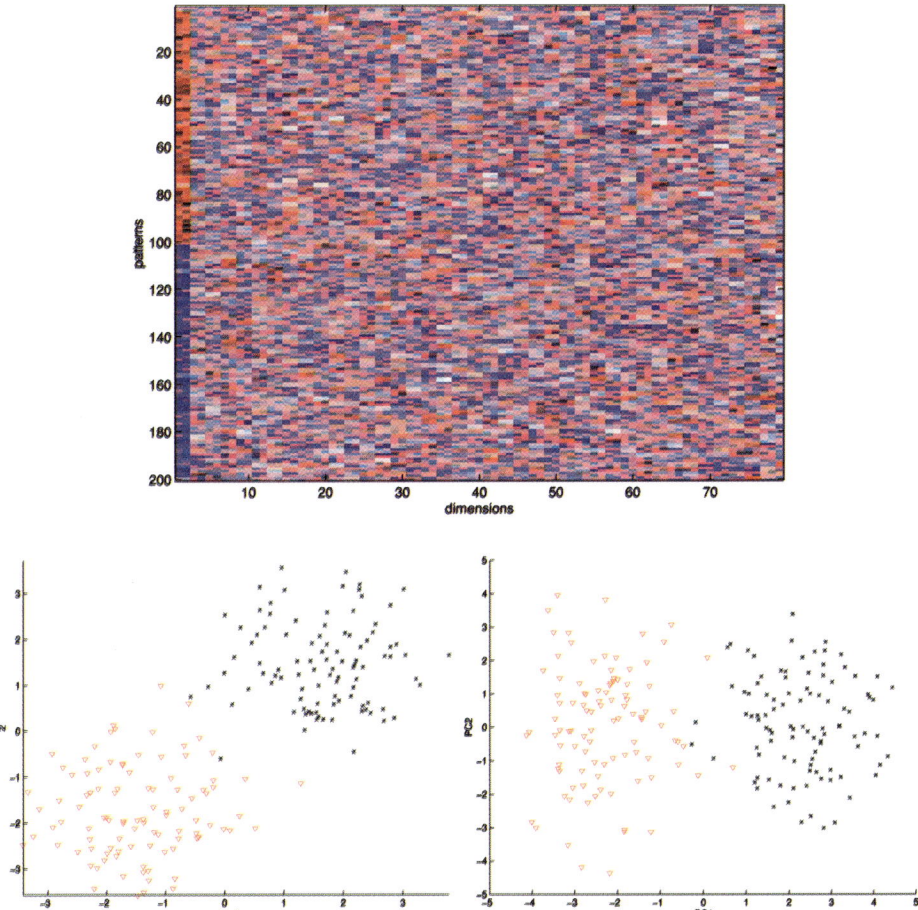

Fig. 1. Synthetic data example. Data consist of 200 patterns with 79 dimensions (variables), with components that are standard Gaussian i.i.d. numbers. As can be seen in the representation of the data (**top**), we added an offset to the first two dimensions (a positive offset of 1.6 for the first 100 patterns and a negative offset of −1.6 for the last 100). This results in the two clusters apparent in the scatter plot of the patterns in the first two dimensions (**bottom left**). The direction that separates the two clusters is captured by the first PC, as shown on the **bottom right** scatter plot of the first two PCs.

Clustering algorithms can be divided into two categories according to the type of *output* they produce:

1. *Hierarchical clustering algorithms*: These output a dendrogram, which is a tree representation of the data whose leaves are the input patterns and whose nonleaf

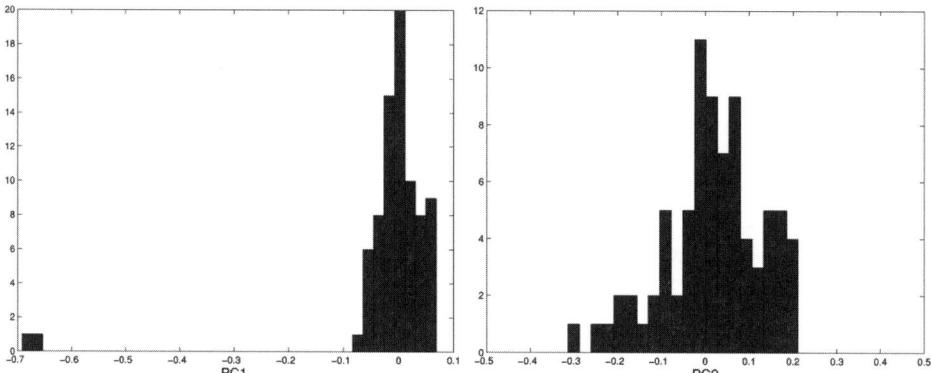

Fig. 2. Histograms of components of first PC (**left**) and second PC (**right**). The count near −0.7 is the value for the first and second variables, representing the 45° rotation seen in **Fig. 1**.

nodes represent a hierarchy of groupings (*see* **Fig. 7**). These come in two flavors: *agglomerative* and *divisive*. Agglomerative algorithms work bottom up, with each pattern in a separate cluster; clusters are then iteratively merged, according to some criterion. Divisive algorithms start from the whole data set in a single cluster and work top down by iteratively dividing each cluster into two components until all clusters are singletons.
2. *Partitional algorithms*: These provide a partition of a data set into a certain number of clusters. Partitional algorithms generally have input parameters that control the number of clusters produced.

A hierarchical clustering algorithm can be used to generate a partition, e.g. by cutting the dendrogram at some level (*see* **Fig. 7** for an illustration). When doing so, we ignore singleton clusters; that is, when cutting the dendrogram to generate k clusters, we look for k non-singleton clusters. We found it useful to impose an even higher threshold, to ignore very small clusters. This approach provides a unified way of considering hierarchical and partitional algorithms, making our methodology applicable to a generic clustering algorithm. We choose to use the average linkage variety of hierarchical clustering *(30,31)*, which has been used extensively in the analysis of gene expression data *(1,2,16)*. In agglomerative hierarchical clustering algorithms, the two nearest (or most similar) clusters are merged at each step. In average linkage clustering, the distance between clusters is defined as the average distance between pairs of patterns that belong to the two clusters; as clusters are merged the distance matrix is updated recursively, making average linkage and other hierarchical clustering algorithms efficient and useful for the large data sets produced in gene expression experiments.

3.1. Clustering Stability

When using a clustering algorithm, several issues must be considered *(10)*:

1. Choice of a clustering algorithm.
2. Choice of a normalization and similarity/dissimilarity measure.
3. Which variables/features to cluster.
4. Which patterns to cluster.
5. Number of clusters: A clustering algorithm provides as output either a partition into k clusters or a hierarchical grouping and does not answer the question of whether there is actually structure in the data, and if there is, what clusters best describe it.

In this section we introduce a framework that helps in making these choices. We use it to choose the number of leading PCs (a form of feature selection) and to compare different types of normalization and similarity measures simultaneously with the discovery of the cluster structure in the data.

The method we are about to describe is based on the following observation: When one looks at two subsamples of a cloud of data patterns with a sampling ratio, f (fraction of patterns sampled) not much smaller than 1 (say $f > 0.5$), one usually observes the same general structure (*see* **Fig. 3**). Thus, it is reasonable to postulate that a partition into k clusters has captured the structure in a data set if partitions into k clusters obtained from running the clustering algorithm with different subsamples are similar.

This idea is implemented as follows: The whole dataset is clustered (the reference clustering); a set of subsamples is generated and clustered as well. For increasing values of k the similarity between partitions of the reference clustering into k clusters and partitions of the subsamples are computed (see the pseudo-code in **Fig. 4**). When the structure in the data is well represented by k clusters, the partition of the reference clustering will be highly similar to partitions of the subsampled data. At a higher value of k, some of the clusters will become unstable, and a broad distribution of similarities will be observed *(12)*. The stable clusters represent the statistically meaningful structure in the data. Lack of structure in the data can also be detected: in this case the transition to instability occurs between $k = 1$ (all partitions identical by definition) and $k = 2$.

The algorithm we have presented has two modular components (not mentioning the clustering algorithm itself): (1) a perturbation of the data set (e.g., subsampling), and (2) a measure of similarity between the perturbed clustering and a reference clustering (or, alternatively, between pairs of perturbed clusterings).

Perturbing the data to probe for stability can be performed in several ways. One can subsample the patterns, as done here when clustering experiments one

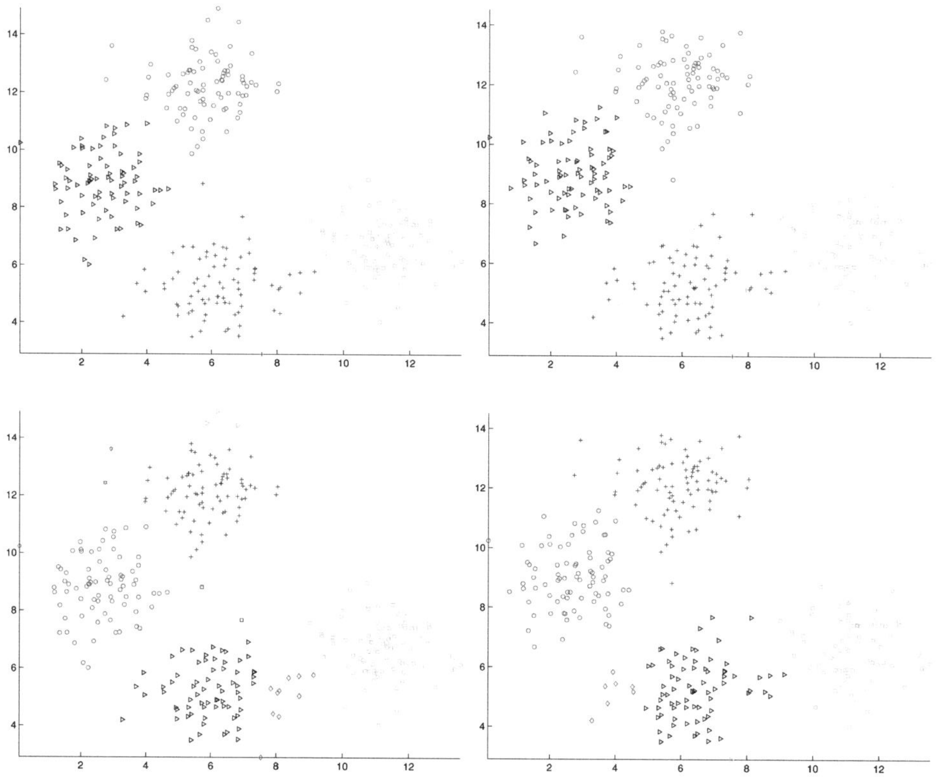

Fig. 3. Two 320-pattern subsamples of a 400-pattern Gaussian mixture. (**Top**) The two subsamples have essentially the same cluster structure, so clustering into four clusters yields similar results. (**Bottom**) The same subsamples; additional clusters can pop up in different locations due to different local substructure in each of the subsamples.

can consider subsampling the genes instead; this is reasonable in view of the redundancy observed in gene expression, since one typically observes many genes that are highly correlated with each other. Another alternative is to add noise to the data *(15)*. In both cases, the user has to decide on the magnitude of the perturbation—what fraction of the features or variables to subsample, or how much noise to add. In our experiments we found that subsampling worked well, and equivalent results were obtained for a wide range of subsampling fractions.

For the second component of the algorithm, a measure of similarity, one can choose one of the several similarity measures proposed in the statistical literature. We introduce a similarity measure originally defined in **ref. *15*** and

Detecting Stable Clusters Using PCA

Input: A dataset X, $k_{\max}$: maximum number of clusters, $num\_subsamples$: number of subsamples.
Output: $S(i, k)$ - a distribution of similarities between partitions into k clusters of a reference clustering and clustering of subsamples; $i = 1, \ldots, num\_subsamples$
Requires: T = cluster(X): A hierarchical clustering algorithm
L = cut-tree(T, k): produces a partition with k non-singleton clusters
$s(L_1, L_2)$: a similarity between two partitions

```
1:  f = 0.8
2:  T = cluster(X)    {the reference clustering}
3:  for i = 1 to num_subsamples do
4:      sub_i = subsamp(X, f)    {sub-sample a fraction f of the data}
5:      T_i = cluster(sub_i)
6:  end for
7:  for k = 2 to k_max do
8:      L_1 = cut-tree(T, k) {partition the reference clustering}
9:      for i = 1 to maximum_iterations do
10:         L_2 = cut-tree(T_i, k)
11:         S(i, k) = s(L_2, L_1) computed only on the patterns of sub_i.
12:     end for
13: end for
```

Fig. 4. Pseudo-code for producing a distribution of similarities. If the distribution of similarities is concentrated near its maximum value, then the corresponding reference partition is said to be stable. In the next subsection it will be refined to assign a stability to individual clusters. Here, it is presented for a hierarchical clustering algorithm, but it can be used with a generic clustering algorithm with minor changes.

then propose an additional one that provides more detailed information about the relationship between the two clusterings.

We define the following matrix representation of a partition:

$$C_{ij} = \begin{cases} 1 & \text{if } g_i \text{ and } g_j \text{ belong to the same cluster and } i \neq j \\ 0 & \text{otherwise} \end{cases} \quad (7)$$

Let two clusterings have matrix representations $C^{(1)}$ and $C^{(2)}$. The dot product

$$\langle C^{(1)}, C^{(2)} \rangle = \sum_{i,j} C_{ij}^{(1)} C_{ij}^{(2)} \quad (8)$$

counts the number of pairs of patterns clustered together in both clusterings and can also be interpreted as the number of edges common to the graphs represented by $C^{(1)}$ and $C^{(2)}$. The dot product satisfies the Cauchy–Schwartz inequality: $\langle C^{(1)}, C^{(2)} \rangle \leq \sqrt{\langle C^{(1)}, C^{(1)} \rangle \langle C^{(2)}, C^{(2)} \rangle}$, and thus can be normalized into a correlation or cosine similarity measure:

$$s(C^{(1)}, C^{(2)}) = \frac{\langle C^{(1)}, C^{(2)} \rangle}{\sqrt{\langle C^{(1)}, C^{(2)} \rangle \langle C^{(1)}, C^{(2)} \rangle}} \qquad (9)$$

Remark. The cluster labels produced by a clustering algorithm ("cluster 1," "cluster 2," etc.) are arbitrary, and the similarity measure just defined is independnt of actual cluster labels: it is defined through the relationship between pairs of patterns—whether patterns i and j belong to the same cluster, regardless of the label given to the cluster.

As mentioned, the signal for the number of clusters can be defined by a transition from highly similar clustering solutions to a wide distribution of similarities. In some cases, the transition is not well defined; if a cluster breaks into two clusters, one large and the other small, the measure of similarity we have presented will still give a high similarity score to the clustering. As a consequence, the number of clusters will be overestimated. To address this issue, and to provide stability scores to individual clusters, we present a new similarity score.

3.2. Associating Clusters of Two Partitions

Here, we represent a partition L by assigning a cluster label from 1 to k to each pattern. The similarity measure defined next is motivated by the success rate from supervised learning, which is the sum of the diagonal elements of the confusion matrix between two sets of labels L_1 and L_2. The confusion matrix measures the size of the intersection between the clusters in two labelings:

$$M_{ij} = |L_1 = i \cap L_2 = j| \qquad (10)$$

in which $|A|$ denotes the cardinality of the set A, and $L = i$ is the set of patterns in cluster i. A confusion matrix like

$$M = \begin{pmatrix} 47 & 2 \\ 1 & 48 \end{pmatrix}$$

represents clusterings that are very similar to each other. The similarity will be quantified by the sum of the diagonal elements. However, in clustering no knowledge about the clusters is assumed, so the labels $1, \ldots, k$ are arbitrary, and any permutation of the labels represents the same clustering. Thus, we might actually get a confusion matrix of the form

$$M = \begin{pmatrix} 1 & 48 \\ 47 & 2 \end{pmatrix}$$

that represents the same clustering. This confusion matrix can be "diagonalized" if we identify cluster 1 in the first labeling with cluster 2 in the second labeling and cluster 2 in the first labeling with cluster 1 in the second labeling. The similarity is then computed as the sum of the diagonal elements of the

Detecting Stable Clusters Using PCA

"diagonalized" confusion matrix. The diagonalization, essentially a permutation of the labels, will be chosen to maximize the similarity. We now formulate these ideas. Given two labelings L_1 and L_2 with k_1, k_2 labels, respectively, we assume $k_1 \leq k_2$. An *association* σ is defined as a one-to-one function σ: $\{1, \ldots, k_1\} \to \{1, \ldots, k_2\}$. The unsupervised analog of the success rate is now defined as follows:

$$s(L_1, L_2) = \max_{\sigma \text{ is an association}} \frac{1}{n} \sum_i M_{i\sigma(i)} \qquad (11)$$

Remark. The computation of the optimal association $\sigma(L_1, L_2)$ by brute-force enumeration is exponential in the number of clusters, since the number of possible one-to-one associations is exponential. To handle this, it was computed by a greedy heuristic: First, clusters from L_1 are associated with the clusters of L_2 with which they have the largest overlap. Conflicts are then resolved one by one; if two clusters are assigned to the same cluster, the one that has smaller overlap with the cluster is assigned to the cluster with which it has the next largest overlap. The process is iterated until no conflicting assigments are present. This method of conflict resolution guarantees the convergence of this process. For small overlap matrices in which exact enumeration can be performed, the results were checked to be identical. Even if, from time to time, the heuristic does not find the optimal solution, our results will not be affected, since we are interested in the statistical properties of the similarity.

Next, we use the optimal association to define concepts of stability for individual patterns and clusters. For a pattern i, two labelings L_1, L_2, and an optimal association σ, define the *patternwise agreement* between L_1 and L_2:

$$\delta_\sigma(i) = \begin{cases} 1 & \sigma(L_1(i)) = L_2(i) \\ 0 & \text{otherwise} \end{cases} \qquad (12)$$

Thus, $\delta_\sigma(i) = 1$ iff pattern i is assigned to the same cluster in the two partitions relative to the association σ. We note that $s(L_1, L_2)$ can be equivalently expressed as $1/n \sum_i \delta_\sigma(i)$.

Now we define *patternwise stability* as the fraction of subsampled partitions in which the subsampled labeling of pattern i agrees with that of the reference labeling, by averaging the patternwise agreement, **Eq. 12**:

$$n(i) = \frac{1}{N_i} \sum_{\text{subsamples}} \delta_\sigma(i) \qquad (13)$$

in which N_i is the number of subsamples in which pattern i appears. The patternwise stability can indicate problem patterns that do not cluster well. Cluster stability is the average of the patternwise stability:

$$c(j) = \frac{1}{|L_1 = j|} \sum_{i \in (L_1 = j)} n(i) \qquad (14)$$

We note that cluster stability should not be interpreted as stability *per se:* Suppose that a stable cluster splits into two clusters in an unstable way, and one cluster is larger than the other. In subsamples of the data, clusters will tend to be associated with the larger subcluster, with the result that the large cluster will have a higher cluster stability. Thus, the stability of the smaller cluster is the one that reflects the instability of this split. This can be seen in **Fig. 7**. Therefore, we define the stability of a reference clustering into k clusters as follows:

$$S_k = \min_j c(j) \qquad (15)$$

In computing S_k we ignore singletons or very small clusters. A dendrogram with stability measurements will be called a *stability annotated dendrogram* (*see* **Fig. 7**).

4. Experiments on Gene Expression Data

In this section, we use the yeast DNA microarray data of Eisen et al. *(1)* as a case study. Functional annotations were used to choose the five functional classes that were most learnable by SVMs *(32)* and noted by Eisen et al. *(1)* to cluster well. We looked at the genes that belong uniquely to these five functional classes. This gave a data set with 208 genes and 79 variables (experiments) in the following classes:

1. Tricarboxylic acid cycle (TCA) (14 genes).
2. Respiration (27 genes).
3. Cytoplasmatic ribosomal proteins (121 genes).
4. Proteasomes (35 genes).
5. Histones (11 genes).

4.1. Clustering of Centered PCA Variables

A scatter plot of the first three centered PCs is shown in **Fig. 5**. Clear cluster structure that corresponds to the functional classes is apparent in the plot. Classes 1 and 2, however, are overlapping. For comparison, we show a scatter plot of the next three PCs (**Fig. 6**), of which visual inspection shows no structure. This supports the premise that cluster structure should be apparent in the leading PCs. To support this premise further, we analyze results of clustering the data in the original variables, and in a few leading PCs. We use the average linkage hierarchical clustering algorithm *(30)* with a Euclidean distance.

When using the first three centered PCs, the functional classes are recovered well by clustering, with the exception of classes 1 and 2 (TCA and respiration),

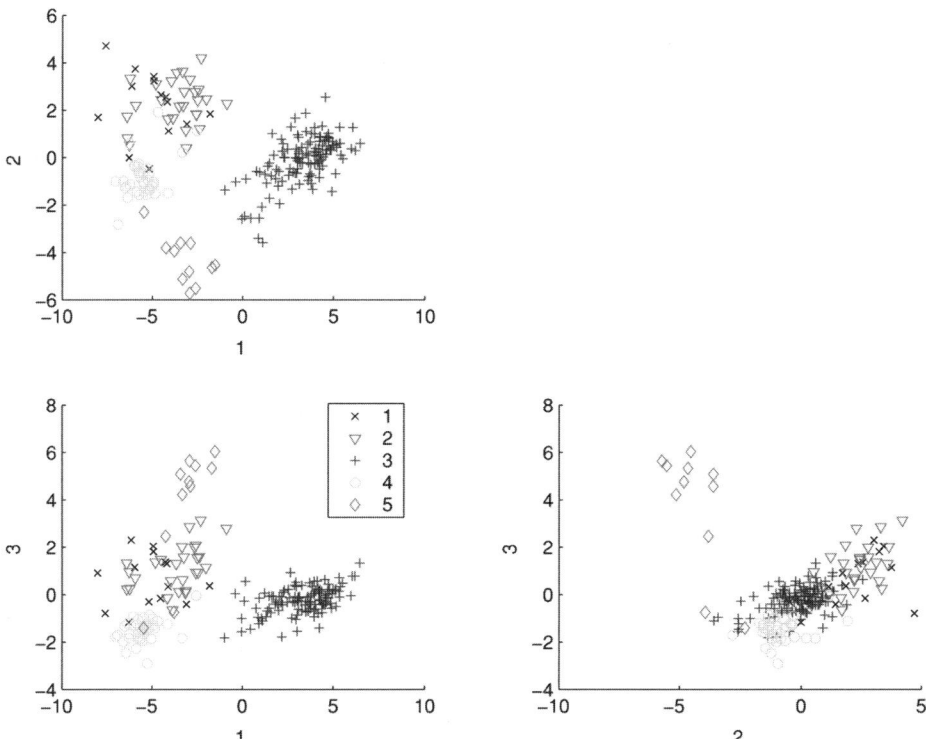

Fig. 5. Scatter plot of first three centered PCs of yeast data. The symbols correspond to the five functional classes.

which cannot be distinguished. The high similarity between the expression patterns of these two classes was also noted by Brown et al. *(32)*. This is seen in the confusion matrix between the functional class labels and the clustering labels for $k = 4$:

		\multicolumn{5}{c}{MIPS classification}				
	classes	1	2	3	4	5
	1	11	27	0	2	0
clustering	2	0	0	121	0	0
	3	2	0	0	33	3
	4	0	0	0	0	8

As already pointed out, when we say that we partition the data into four clusters we mean four nonsingleton clusters (or, more generally, clusters larger than some threshold). This allows us to ignore outliers, making comparisons of

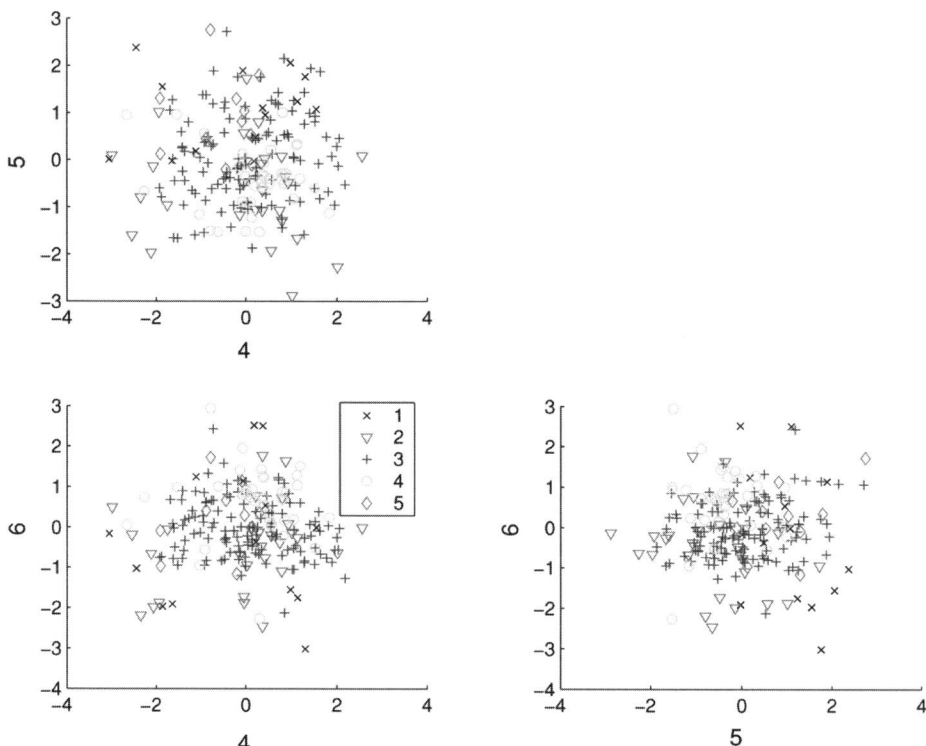

Fig. 6. Scatter plot of PCs four to six for yeast data.

partitions into k clusters more meaningful, since an outlier will not appear in all the subsamples clustered. Indeed, in the dendrogram in **Fig. 7** for $k = 4$, we find one singleton cluster for a total of five clusters.

The choice of $k = 4$ is justified by the stability method: The top dendrogram in **Fig. 7** shows a stability annotated dendrogram for the PCA data. The numbers at each node indicate the cluster stability of the corresponding cluster averaged over all the levels at which the cluster appears. All the functional classes appear in clusters with cluster stability above 0.96. The ribosomal cluster then splits into two clusters, one of them having a stability value of 0.62, indicating that this split is unstable; the stability annotated dendrogram (**Fig. 7**) and the plot of the minimum cluster stability, S_k (**Fig. 9**), show that the functional classes correspond to the stable clusters in the data.

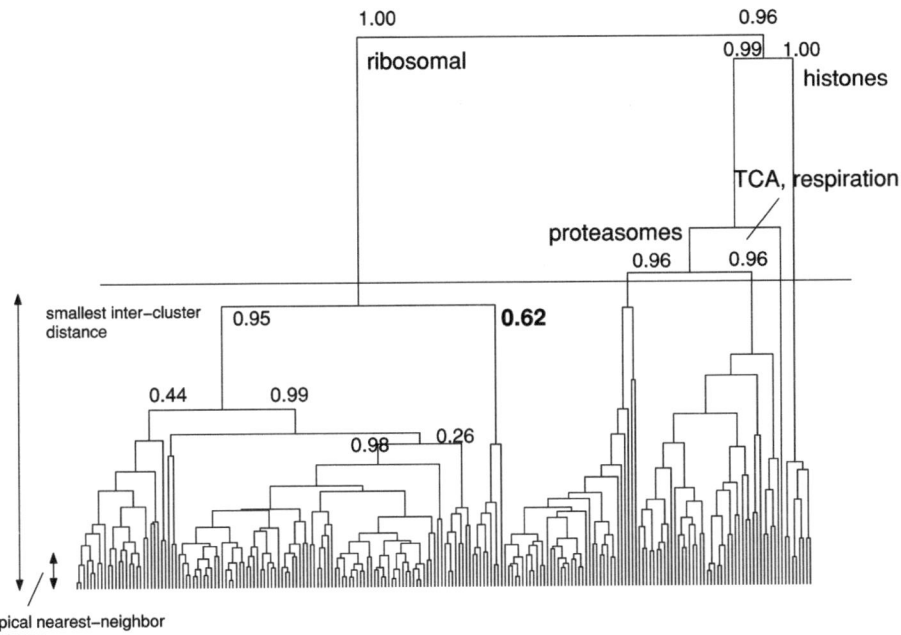

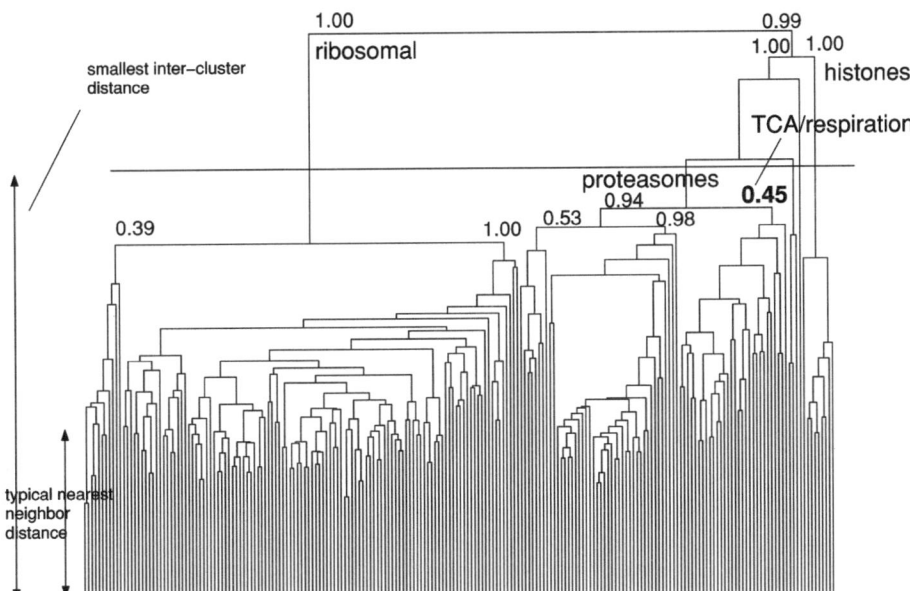

Fig. 7. Stability annotated dendrograms for yeast data. Numbers represent the cluster stability of a node. Cluster stability is averaged over all the levels in the hierarchy in which a cluster appears. The horizontal line represents the cutoff suggested by cluster stability; boldface represents the stability of the corresponding unstable split. (**Top**) Data composed of three leading centered PCA variables. The vertical axis gives intercluster distances. The nodes that correspond to the functional classes are indicated. (**Bottom**) Clustering of all 79 centered variables.

When using all the variables (or, equivalently, all PCs), the functional classes are recovered at $k = 5$, but not as well:

	classes	MIPS classification				
		1	2	3	4	5
clustering	1	6	22	0	2	0
	2	3	0	0	0	0
	3	0	0	121	0	0
	4	4	5	0	33	2
	5	0	0	0	0	9

The same confusion matrix was observed when using 20 leading PCs or more. However, clustering into $k = 5$ clusters is not justified by the stability criterion: the split into the proteasome and TCA/respiration clusters is highly unstable (see the bold number in the bottom stability annotated dendrogram in **Fig. 7**). Only a partition into three clusters is stable.

Inspecting the dendrograms also reveals an important property of clustering of the leading PCs: the cluster structure is more apparent in the PCA dendrogram, using $q = 3$ components. Using all the variables, the distances between nearest neighbors and the distances between clusters are comparable, whereas for $q = 3$ nearest-neighbor distances are very small compared with the distances between clusters, making the clusters more well defined and, consequently, more stable.

Using the comparison with the "true" labels, it was clear that the three-PC data provided more detailed stable structure (three vs. four clusters). "True" labels are not always available, so we would like a method for telling which of two partitions is more "refined." We define a *refinement score*, also defined using the confusion matrix:

$$r(L_1, L_2) = \frac{1}{n} \sum_i \max_j M_{ij} \qquad (16)$$

The motivation for this score is illustrated with the help of **Fig. 8**. Both blue clusters have a big overlap with the large red clusters and $r(blue, red) = 1/16$ $(7 + 7)$, which is close to 1, in agreement with our intuition that the blue clustering is basically a division of the big red cluster into two clusters. On the other hand $r(red, blue) = 1/16$ $(2 + 7)$, with a much lower score. It is straightforward to verify that $1/2 \leq r(L_1, L_2) \leq 1$, and thus $r(red, blue)$ is close to its lower bound. To make the relationship with the previous score more clear, we note that it can be defined as $s(L_1, L_2)$, in which the association σ is not constrained to be one to one, so that each cluster is associated with the cluster

Detecting Stable Clusters Using PCA

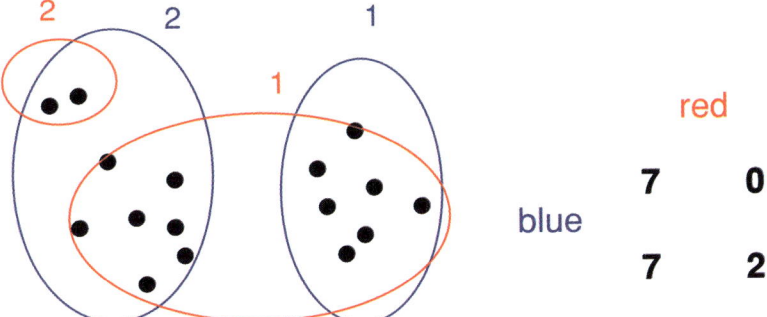

Fig. 8. Two clusterings of a data set, represented by red and blue, with associated confusion matrix.

with which it has the maximum overlap. If L_1 is obtained by splitting one of the clusters of L_2, then $r(L_1, L_2) = 1$, and $r(L_2, L_1) = 1$ – the fraction of patterns in the smaller of the two clusters that have split. The refinement score is interesting when it comes to comparing clusterings obtained from different algorithms, different subsets of features, or different dendrograms obtained with any kind of parameter change. Given two stable partitions L_1 and L_2, one can then determine which of the two partitions is more refined according to which of $r(L_1, L_2)$ or $r(L_2, L_1)$ is larger. Merely counting the number of clusters to estimate refinement can be misleading since given two partitions with an identical number of clusters, one may be more refined than the other (as exemplified in **Fig. 7**). It is even possible that a partition with a smaller number of clusters will be more refined than a partition with a larger number of clusters.

4.2. Stability-Based Choice of Number of PCs

The results of the previous section indicated that three PCs gave a more stable clustering than that obtained using all the variables. Next, we will determine the best number of PCs. In the choice of PCs we will restrict ourselves to choosing the number of leading components, rather than choosing the best components, not necessarily by order of variance. We still consider centered-PCA data.

The stability of the clustering as measured by the minimum cluster stability, S_k, is plotted for a varying number of PCs in **Fig. 9**. Partitions into up to three clusters are stable regardless of the number of PCs, as evidenced by S_k being close to 1. Partitions into four clusters are most stable for three or four PCs,

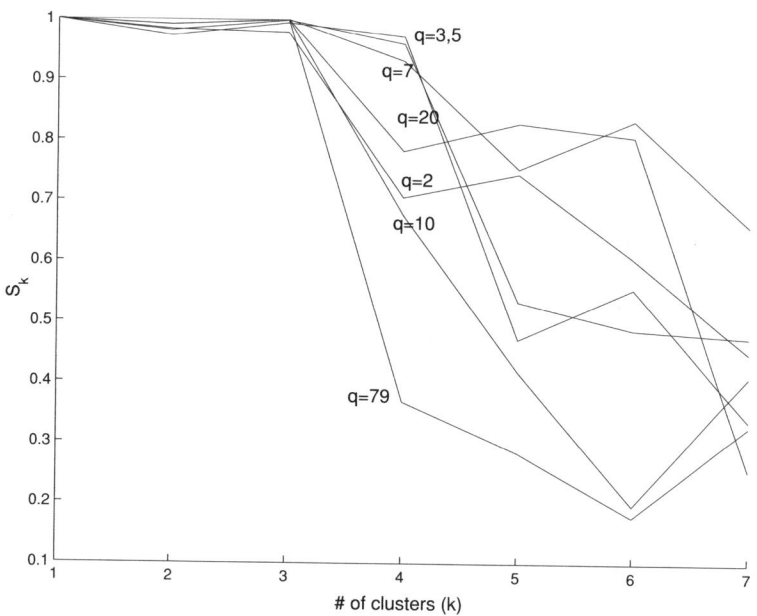

Fig. 9. For each k, the average of the minimum cluster stability, S_k (cf. Eq. 15), is estimated for a varying number of PCs (q).

and slightly less so for seven PCs, with S_k still close to 1. For a higher number of PCs, the stability of four clusters becomes lower (between 0.4 and 0.8). Moreover, the stable structure observed in three to seven components at $k = 4$ is only observed at $k = 5$ for a higher number of PCs (and is unstable). The cluster structure is less stable with two PCs than in three to seven. The scatter plot of the PCs shows that the third PC still contains relevant structure; this is in agreement with the instability for two components, which are not sufficient.

To conclude, a small number of leading PCs produced clustering solutions that were more stable (i.e., significant) and also agreed better with the known classification.

Our observation on the power of PCA to produce a stable classification was seen to hold on data sets in other domains as well. Typically, when the number of clusters was higher, more PCs were required to capture the cluster structure. Other clustering algorithms that use the Euclidean distance were also seen to benefit from a preprocessing using PCA *(33)*.

4.3. Clustering of Standardized PCs

In this section, we compare clustering results on centered and standardized PCs. We limit ourselves to the comparison of $q = 3$ and $q = 79$ PCs. A scatter

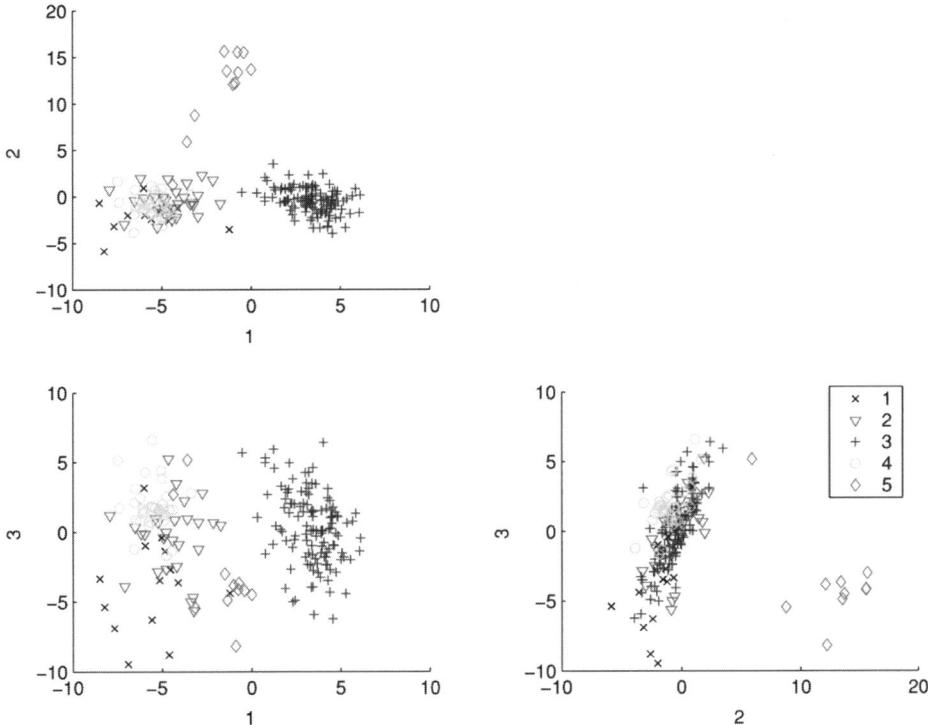

Fig. 10. Scatter plot of first three standardized PCs.

plot of the first three standardized PCs is shown in **Fig. 10**. These show less cluster structure than in the centered PCs (**Fig. 5**). The stability validation tool indicates the existence of four clusters, with the following confusion matrix:

	classes	MIPS classification				
		1	2	3	4	5
clustering	1	9	27	0	35	2
	2	5	0	0	0	0
	3	0	0	121	0	0
	4	0	0	0	0	9

In this case, classes 1, 2, and 4 cannot be distinguished by clustering. A similar confusion matrix is obtained when clustering the standardized variables without applying PCA. We conclude that the degradation in the recovery of the functional classes should be attributed to normalization, rather than to the use of PCA. Other investigators have also found that standardization can deteriorate the quality of clustering *(29,34)*.

Let $L_{centered}$ and $L_{standardized}$ be the stable partitions into four and three clusters, respectively, of the centered and standardized PCA data using three PCs. The refinement scores for these partitions are found to be $r(L_{centered}, L_{standardized}) = 0.8125$ and $r(L_{standardized}, L_{centered}) = 0.995$, showing that the centered clustering contains more detailed stable structure. Thus, the known cluster labels are not necessary to arrive at this conclusion.

4.4. Clustering Using Pearson Correlation

Here we report results using the Pearson correlation similarity measure (the gene–gene correlation matrix). Clustering using the Pearson correlation as a similarity measure recovered the functional classes in a stable way without the use of PCA (see the stability annotated dendrogram in **Fig. 11**). It seems that the Pearson correlation is less susceptible to noise than the Euclidean distance: the Pearson correlation clustering is more stable than the Euclidean clustering that uses all the variables; also compare the nearest-neighbor distances, which are much larger in the Euclidean case. This may be the result of the Pearson correlation similarity being a sum of terms that are either positive or negative, resulting in some of the noise canceling out; in the case of the Euclidean distance, no cancelation can occur since all terms are positive. The Pearson correlation did not benefit from the use of PCA.

5. Other Methods for Choosing PCs

In a recent article, it was claimed that PCA does not generally improve the quality of clustering of gene expression data *(19)*. We suspect that the claim was a result of the use of standardization as a normalization, which in our analysis reduced the quality of the clustering, rather than the use of PCA. The authors' criterion for choosing components was external—comparison with known labels—and thus cannot be used in general. We use a criterion that does not require external validation. However, running it is relatively time-consuming since it requires running the clustering algorithm a large number of times to estimate the stability (a value of 100 was used here). Therefore, we restricted the feature selection to the choice of the number of leading PCs.

When using PCA to approximate a data matrix, the fraction of the total variance in the leading PCs is used as a criterion for choosing how many of them to use *(20)*. In the yeast data analyzed in this chapter, the first three PCs contain only 31% of the total variance in the data. Yet, 3–5 PCs out of 79 provide the most stable clustering, which also agrees best with the known labels. Thus, the total variance is a poor way of choosing the number of PCs when the objective is clustering. On the contrary, we do not wish to reconstruct the matrix, only the essential features that are responsible for cluster structure.

Detecting Stable Clusters Using PCA

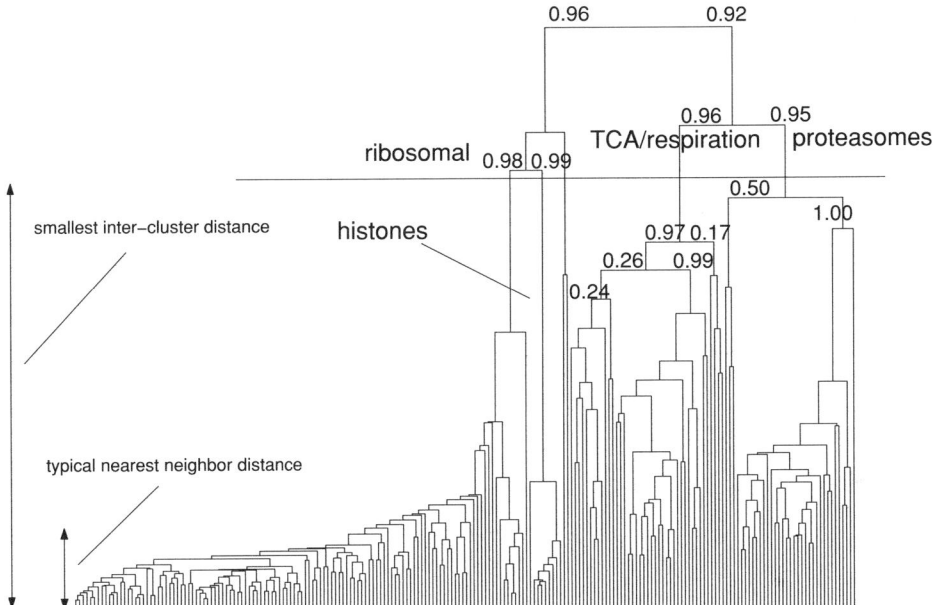

Fig. 11. Stability annotated dendrogram for yeast data with the Pearson correlation similarity measure. The vertical axis is 1–correlation. The functional classes are stable; their subclusters are unstable.

6. Conclusion

In this chapter, we proposed a novel methodology for evaluating the merit of clustering solutions. It is based on the premise that well-defined cluster structure should be stable under perturbation of the data. We introduced notions of stability at the level of a partition, cluster, and pattern that allow us, in particular, to generate stability annotated dendrograms. Using stability as a figure of merit, along with a measure of cluster refinement, allows us to compare stable solutions run using different normalizations, similarity measures, clustering algorithms, or input features.

We demonstrated this methodology on the task of choosing the number of PCs and normalization to be used to cluster a yeast gene expression data set. It was shown that PCA improves the extraction of cluster structure, with respect to stability, refinement, and coincidence with known "ground truth" labels. This means that, in this data set, the cluster structure is present in directions of largest variance (or best data reconstruction). Beyond this simple example, our methodology can be used for many "model selection" problems in clustering, including selecting the clustering algorithm itself, the parameters of the algorithm, the variables and patterns to cluster, the similarity, normalization or other preprocessing, and the number of clusters. It allows us not only to compare clustering solutions, but also to detect the presence or absence of structure in data. In our analysis, we used a hierarchical clustering algorithm, but any other clustering algorithm can be used. The versatility and universality of our methodology, combined with its simplicity, may appeal to practitioners in bioinformatics and other fields. Further work includes laying the theoretical foundations of the methodology and additional testing of the ideas in other domains.

Acknowledgments

We thank André Elisseeff and Bronwyn Eisenberg for their helpful comments on the manuscript.

References

1. Eisen, M., Spellman, P., Brown, P., and Botstein, D. (1998) Cluster analysis and display of genome-wide expression patterns. *Proc. Natl. Acad. Sci. USA* **95**, 14,863–14,868.
2. Quackenbush, J. (2001) Computational anaysis of microarray data. *Nat. Rev. Genet.* **2**, 418–427.
3. Ross, D., Scherf, U., Eisen, M., et al. (2000) Systematic variation in gene expression patterns in human cancer cell lines. *Nat. Genet.* **24**, 227–235.
4. D'haeseleer, P., Liang, S., and Somogyi, R. (2000) Genetic network inference: from co-expression clustering to reverse engineering. *Bioinformatics* **16**, 707–726.
5. Alon, U., Barkai, N., Notterman, D., Gish, G., Ybarra, S., Mack, D., and Levine, A. J. (1999) Broad patterns of gene expression revealed by clustering analysis of tumor and normal colon tissues probed by oligonucleotide arrays. *PNAS* **96**, 6745–6750.
6. Alizadeh, A. A., et al. (2000) Distinct types of diffuse large B-cell lymphoma identified by gene expression profiling. *Nature* **403**, 503–511.
7. Scherf, U., Ross, D. T., Waltham, M., et al. (2000) A gene expression database for the molecular pharmacology of cancer. *Nat. Genet.* **24**, 236–244.
8. Getz, G., Levine, E., and Domany, E. (2000) Coupled two-way clustering analysis of gene microarray data. *Proc. Natl. Acad. Sci. USA* **94**, 12,079–12,084.
9. Shamir, R. and Sharan, R. (2001) Algorithmic approaches to clustering gene expression data, in *Current Topics in Computational Biology* (Jiang, T., Smith, T., Xu, Y., and Zhang, M., eds.), MIT Press, Cambridge, MA, pp. 269–299.
10. Milligan, G. (1996) Clustering validation: results and implications for applied analysis, in *Clustering and Classification* (Arabie, P., Hubert, L., and Soete, G. D., eds.), World Scientific, River Edge, NJ, pp. 341–374.
11. Tibshirani, R., Walther, G., and Hastie, T. (2001) Estimating the number of clusters in a dataset via the gap statistic. *J. Roy. Stat. Soc.* **63**, 411–423.

12. Ben-Hur, A., Elisseeff, A., and Guyon, I. (2002) A stability based method for discovering structure in clustered data, in *Pacific Symposium on Biocomputing* (Altman, R., Dunker, A., Hunter, L., Lauderdale, K., and Klein, T., eds.), World Scientific, River Edge, NJ, pp. 6–17.
13. Levine, E. and Domany, E. (2001) Resampling method for unsupervised estimation of cluster validity. *Neural Comp.* **13,** 2573–2593.
14. Fridlyand, J. (2001) Resampling methods for variable selection and classification: applications to genomics, PhD thesis, University of California at Berkeley.
15. Fowlkes, E. and Mallows, C. (1983) A method for comparing two hierarchical clusterings. *J. Am. Stat. Assoc.* **78,** 553–584.
16. Bittner, M., Meltzer, P., Chen, Y., et al. (2000) Molecular classification of cutaneous malignant melanoma by gene expression profiling. *Nature* **406,** 536–540.
17. Yeung, K., Haynor, D., and Ruzzo, W. (2001) Validating clustering for gene expression data. *Bioinformatics* **17,** 309–318.
18. Milligan, G. (1980) An examination of the effect of six types of error perturbation on fifteen clustering algorithms. *Psychometrika* **45,** 325–342.
19. Yeung, K. and Ruzzo, W. (2001) An empirical study of principal component analysis for clustering gene expression data. *Bioinformatics* **17,** 763–774.
20. Jackson, J. (1991) *A User's Guide to Principal Components*, John Wiley & Sons, New York, NY.
21. Jolliffe, I. (1986) *Principal Component Analysis*, Springer-Verlag, New York, NY.
22. Jolliffe, I. (1972) Discarding variables in principal component analysis I: artificial data. *Appl. Stat.* **21,** 160–173.
23. Jolliffe, I. (1972) Discarding variables in principal component analysis II: real data. *Appl. Stat.* **21,** 160–173.
24. Hastie, T., Tibshirani, R., Eisen, M., Alizadeh, A., Levy, R., Staudt, L., Chan, W., Botstein, D., and Brown, P. (2000) "Gene shaving" as a method for identifying distinct sets of genes with similar expression patterns. *Genome Biol.* **1,** 1–21.
25. Alter, O., Brown, P., and Botstein, D. (2000) Singular value decomposition for genomewide expression data processing and modeling. *Proc. Natl. Acad. Sci. USA* **97,** 10101–10106.
26. Raychaudhuri, S., Stuart, J., and Altman, R. (2000) Principal components analysis to summarize microarray experiments: application to sporulation time series. *Pacific Symp. Biocomp.* **5,** 452–463.
27. Hilsenbeck, S., Friedrichs, W., Schiff, R., O'Connell, P., Hansen, R., Osborne, C., and Fuqua, S. (1999) Statistical analysis of array expression data as applied to the problem of tamoxifen resistance. *J. Natl. Cancer Inst.* **91,** 453–459.
28. Wen, X., Fuhrman, S., Michaels, G., Carr, D., Smith, S., Barker, J., and Somogyi, R. (1998) Large-scale temporal gene expression mapping of central nervous system development. *Proc. Natl. Acad. Sci. USA* **95,** 334–339.
29. Milligan, G. and Cooper, M. (1988) A study of variable standardization. *J. Classif.* **5,** 181–204.

30. Jain, A. and Dubes, R. (1988) *Algorithms for Clustering Data*, Prentice Hall, Englewood Cliffs, NJ.
31. Kaufman, L. and Rousseeuw, P. (1990) *Finding Groups in Data*, Wiley Interscience, John Wiley & Sons, New York, NY.
32. Brown, M., Grundy, W., Lin, D., Cristianini, N., Sugnet, C., Ares, M., and Haussler, D. (2000) Knowledge-based analysis of microarray gene expression data by using support vector machines. *Proc. Natl. Acad. Sci. USA* **97,** 262–267. www.cse.ucsc.edu/research/compbio/genex.
33. Ben-Hur, A., Horn, D., Siegelmann, H., and Vapnik, V. (2001) Support vector clustering. *J. Machine Learn. Res.* **2,** 125–137.
34. Anderberg, M. (1983) *Cluster Analysis for Applications*, Academic, New York.

13

Clustering in Life Sciences

Ying Zhao and George Karypis

1. Introduction

Clustering is the task of organizing a set of objects into meaningful groups. These groups can be disjoint, overlapping, or organized in some hierarchical fashion. The key element of clustering is the notion that the discovered groups are *meaningful*. This definition is intentionally vague, because what constitutes meaningful is, to a large extent, application dependent. In some applications, this may translate to groups in which the pairwise similarity between their objects is maximized, and the pairwise similarity between objects of different groups is minimized. In some other applications, this may translate to groups that contain objects that share some key characteristics, even though their overall similarity is not the highest. Clustering is an exploratory tool for analyzing large data sets and has been used extensively in numerous application areas.

Clustering has a wide range of applications in life sciences and, over the years, has been used in many areas including the analysis of clinical information, phylogeny, genomics, and proteomics. For example, clustering algorithms applied to gene expression data can be used to identify coregulated genes and provide a genetic fingerprint for various diseases. Clustering algorithms applied on the entire database of known proteins can be used to automatically organize the different proteins into close- and distant-related families, and to identify subsequences that are mostly preserved across proteins *(1–5)*. Similarly, clustering algorithms applied to the tertiary structural data sets can be used to perform a similar organization and provide insights into the rate of change between sequence and structure *(6,7)*.

The primary goal of this chapter is to provide an overview of the various issues involved in clustering large data sets, describe the merits and underlying assumptions of some of the commonly used clustering approaches, and provide insights on how to cluster data sets arising in various areas within life sciences. Toward this end, the chapter is organized broadly in three parts. The first part (**Subheadings 2.–4.**) describes the various types of clustering algorithms developed over the years, the various methods for computing the similarity between objects arising in life sciences, and methods for assessing the quality of the clusters. The second part (**Subheading 5.**) focuses on the problem of clustering data arising from microarray experiments and describes some of the commonly used approaches. Finally, the third part (**Subheading 6.**) provides a brief introduction to CLUTO, a general purpose toolkit for clustering various data sets, with an emphasis on its applications to problems and analysis requirements within life sciences.

2. Types of Clustering Algorithms

The topic of clustering has been extensively studied in many scientific disciplines and a variety of different algorithms have been developed *(8–20)*. Two recent surveys on the topic *(21,22)* offer a comprehensive summary of the different applications and algorithms. These algorithms can be categorized along different dimensions based on either the underlying methodology of the algorithm, leading to *partitional* or *agglomerative* approaches; the structure of the final solution, leading to *hierarchical* or *nonhierarchical* solutions; the characteristics of the space in which they operate, leading to *feature* or *similarity* approaches; or the type of clusters that they discover, leading to *globular* or *transitive* clustering methods.

2.1. Agglomerative and Partitional Algorithms

Partitional algorithms, such as *K*-means *(9,23)*, *K*-medoids *(9,11,23)*, probabilistic *(10,24)*, graph-partitioning based *(9,25–27)*, or spectral based *(28)*, find the clusters by partitioning the entire data set into either a predetermined or an automatically derived number of clusters.

Partitional clustering algorithms compute a k-way clustering of a set of objects either directly or via a sequence of repeated bisections. A direct k-way clustering is commonly computed as follows. Initially, a set of k objects is selected from the data sets to act as the *seeds* of the k clusters. Then, for each object, its similarity to these k seeds is computed, and it is assigned to the cluster corresponding to its most similar seed. This forms the initial k-way clustering. This clustering is then repeatedly refined so that it optimizes a desired clustering criterion function. A k-way partitioning via repeated bisections is obtained by recursively applying the above algorithm to compute two-way clustering (i.e., bisections). Initially, the objects are partitioned into two clusters, then one of these clusters is selected and is further bisected, and so on. This process continues $k - 1$ times, leading to k clusters. Each of these bisections is performed so that the resulting two-way clustering solution optimizes a particular criterion function.

Criterion functions used in the partitional clustering reflect the underlying definition of the "goodness" of clusters. The partitional clustering can be considered an optimization procedure that tries to create high-quality clusters according to a particular criterion function. Many criterion functions have been proposed *(9,29,30)*; and some are described in **Subheading 6.** Criterion functions measure various aspects of intracluster similarity, intercluster dissimilarity, and their combinations. These criterion functions utilize different views of the underlying collection, either by modeling the objects as vectors in a high-dimensional space or by modeling the collection as a graph.

Hierarchical agglomerative algorithms find the clusters by initially assigning each object to its own cluster and then repeatedly merging pairs of clusters until a certain stopping criterion is met. Consider an n-object data set and the clustering solution that has been computed after performing l merging steps. This solution will contain exactly $n - l$ clusters, as each merging step reduces the number of clusters by one. Now, given this $(n - l)$-way clustering solution, the pair of clusters that is selected to be merged next, is the one that leads to an $(n - l - 1)$-way solution that optimizes a particular criterion function. That is, each one of the $(n - l) \times (n - l - 1)/2$ pairs of possible merges is evaluated, and the one that leads to a clustering solution that has the maximum (or minimum) value of the particular criterion function is selected. Thus, the criterion function is *locally* optimized within each particular stage of agglomerative algorithms. Depending on the desired solution, this process continues either until there are only k clusters left, or when the entire agglomerative tree has been obtained.

The three basic criteria to determine which pair of clusters to be merged next are single link *(31)*, complete -link *(32)*, and group average Unweighted Pair Group Method with Arithmetic Mean (UPGMA) *(9)*. The single-link criterion function measures the similarity of two clusters by the maximum similarity between any pair of objects from each cluster, whereas the complete-link criterion uses the minimum similarity. In general, both the single- and the complete-link approaches do not work very well because they either base their decisions on a limited amount of information (single link), or assume that all the objects in the cluster are very similar to each other (complete link). On the other hand, the group average approach measures the similarity of two clusters by the average of the pairwise similarity of the objects from each cluster and does not suffer from the problems arising with the single- and complete-link approaches. In addition to these three basic approaches, a number of more sophisticated schemes have been developed, such as CURE *(18)*, ROCK *(19)*, CHAMELEON *(20)*, that have been shown to produce superior results.

Finally, hierarchical algorithms produce a clustering solution that forms a dendrogram, with a single all-inclusive cluster at the top and single-point clusters at the leaves. By contrast, in nonhierarchical algorithms, there tends to be no relation between the clustering solutions produced at different levels of granularity.

2.2. Feature- and Similarity-Based Clustering Algorithms

Another distinction between the different clustering algorithms is whether or not they operate on the object's feature space or on a derived similarity (or distance) space. K-means-based algorithms are the prototypical examples of methods that operate on the original feature space. In this class of algorithms, each object is represented as a multidimensional feature vector, and the clustering solution is obtained by iteratively optimizing the similarity (or distance) between each object and its cluster centroid. By contrast, similarity-based algorithms compute the clustering solution by first computing the pairwise similarities between all the objects and then use these similarities to drive the overall clustering solution. Hierarchical agglomerative schemes, graph-based schemes, as well as K-medoid, fall into this category. The advantages of similarity-based methods is that they can be used to cluster a wide variety of data sets, provided that reasonable methods exist for computing the pairwise similarity between objects. For this reason, they have been used to cluster both sequential *(1,2)* and graph data sets *(33,34)*, especially in biological applications. However, there has been limited work in developing clustering algorithms that operate directly on the sequence or graph datasets *(35)*.

Similarity-based approaches do have two key limitations. First, their computational requirements are high since they have to compute the pairwise similarity between all the objects that need to be clustered. As a result, such algorithms can only be applied to relatively small data sets (a few thousand objects), and they cannot be effectively used to cluster the data sets arising in many fields within life sciences. Second, by not using the object's feature space and relying only on the pairwise similarities, they tend to produce suboptimal clustering solutions especially when the similarities are low relative to the cluster sizes. The key reason for this is that these algorithms can only determine the overall similarity of a collection of objects (i.e., a cluster) by using measures derived from the pairwise similarities (e.g., average, median, or minimum pairwise similarities). However, such measures, unless the overall similarity between the members of different clusters is high, are quite unreliable because they cannot capture what is *common* between the different objects in the collection.

Clustering algorithms that operate in the object's feature space can overcome both of these limitations. Since they do not require the precomputation of the pairwise similarities, fast partitional algorithms can be used to find the clusters, and since their clustering decisions are made in the object's feature space, they can potentially lead to better clusters by correctly evaluating the similarity between a collection of objects. For example, in the context of clustering protein sequences, the proteins in each cluster can be analyzed to determine the conserved blocks, and use only these blocks in computing the similarity between the sequences (an idea formalized by profile HMM approaches *(36,37)*). Recent studies in the context of clustering large high-dimensional data sets done by various groups *(38–41)* show the advantages of such algorithms over those based on similarity.

2.3. Globular and Transitive Clustering Algorithms

Besides the operational differences among various clustering algorithms, another key distinction among them is the type of clusters that they discover. There are two general types of clusters that often arise in different application domains. What differentiates these types is the relationship between the cluster's objects and the dimensions of their feature space.

The first type of cluster contains objects that exhibit a strong pattern of conservation along a subset of their dimensions. That is, there is a subset of the original dimensions in which a large fraction of the objects agree. For example, if the dimensions correspond to different protein motifs, then a collection of proteins will form a cluster, if a subset of motifs is present in a large fraction of the proteins. This subset of dimensions is often referred to as a *subspace*, and the aforementioned property can be viewed as the cluster's objects and its associated dimensions forming a *dense subspace*. Of course, the number of dimensions in these dense subspaces, and the density (i.e., how large the *fraction* is of the objects that share the same dimensions) will be different from cluster to cluster. Exactly this variation in subspace size and density (and the fact that an object can be part of multiple disjoint or overlapping dense subspaces) is what complicates the problem of discovering this type of clusters. There are a number of application areas in which such clusters give rise to meaningful groupings of the objects (i.e., domain experts will tend to agree that the clusters are correct). Such areas includes clustering documents based on the terms that they contain, clustering customers based on the products they purchase, clustering genes based on their expression levels, and clustering proteins based on the motifs that they contain.

The second type of cluster contains objects in which, again, there exists a subspace associated with that cluster. However, unlike the earlier case, in these clusters there will be subclusters that may share a very small number of the subspace's dimension, but there will be *a strong path* within that cluster that will connect them. By "strong path" we mean that if *A* and *B* are two subclusters that share only a few dimensions, then there will be another set of subclusters $X_1, X_2, \ldots, X_k$, that belong to the cluster, such that each of the subcluster pairs $(A, X_1), (X_1, X_2), \ldots, (X_k, B)$ will share many of the subspace's dimensions. What complicates cluster discovery in this setting is that the connections (i.e., shared subspace dimensions) between subclusters within a particular cluster will tend to be of different strength. Examples of this type of cluster include protein clusters with distant homologies or clusters of points that form spatially contiguous regions.

Our discussion so far has focused on the relationship between the objects and their feature space. However, these two classes of clusters can also be understood in terms of the object-to-object similarity graph. The first type of clusters will tend to contain objects in which the similarity among all pairs of objects will be high. By contrast, in the second type of cluster, there will be a lot of objects whose direct pairwise similarity will be quite low, but these objects will be connected by many paths that stay within the cluster that traverse high-similarity edges. The names of these two cluster types were inspired by this similarity-based view, and they are referred to as *globular* and *transitive* clusters, respectively.

The various clustering algorithms are generally suited for finding either globular or transitive clusters. In general, clustering-criterion-driven partitional clustering algorithms such as K-means and its variants and agglomerative algorithms using the complete-link or the group-average method are suited for finding globular clusters. On the other hand, the single-link method of the agglomerative algorithm, and graph-partitioning-based clustering algorithms that operate on a nearest-neighbor similarity graph are suited for finding transitive clusters. Finally, specialized algorithms, called *subspace* clustering methods, have been developed for explicitly finding either globular or transitive clusters by operating directly in the object's feature space *(42–44)*.

3. Methods for Measuring Similarity Between Objects

In general, the method used to compute the similarity between two objects depends on two factors. The first factor has to do with how the objects are actually being represented. For example, the method to measure similarity between two objects represented by a set of attribute-value pairs will be entirely different from the method used to compute the similarity between two DNA sequences or two three-dimensional (3D) protein structures. The second factor is much more subjective and deals with the actual goal of clustering. Different analysis requirements may give rise to entirely different similarity measures and different clustering solutions. This section focuses on discussing various methods for computing the similarity between objects that address both of these factors.

The diverse nature of biological sciences and the shear complexity of the underlying physicochemical and evolutionary principles that need to be modeled give rise to numerous clustering problems involving a wide range of different objects. The most prominent of them are the following:

1. *Multidimensional vectors*: Each object is represented by a set of attribute-value pairs. The meaning of the attributes (also referred to as variables or features) is application dependent and includes data sets like those arising from various measurements (e.g., gene expression data), or from various clinical sources (e.g., drug response, disease states).
2. *Sequences*: Each object is represented as a sequence of symbols or events. The meaning of these symbols or events also depends on the underlying application and includes objects such as DNA and protein sequences, sequences of secondary structure elements, temporal measurements of various quantities such as gene expressions, and historical observations of disease states.
3. *Structures*: Each object is represented as a two-dimensional (2D) or 3D structure. The primary examples of such data sets include the spatial distribution of various quantities of interest within various cells, and the 3D geometry of chemical molecules such as enzymes and proteins.

The following sections describe some of the most popular methods for computing the similarity for all these types of objects.

3.1. Similarity Between Multidimensional Objects

There are a variety of methods for computing the similarity between two objects that are represented by a set of attribute-value pairs. These methods, to a large extent, depend on the nature of the attributes themselves and the characteristics of the objects that we need to model by the similarity function.

From the point of similarity calculations, there are two general types of attributes. The first consists of attributes whose range of values are *continuous*. This includes both integer- and real-valued variables, as well as attributes whose allowed set of values are thought to be part of an ordered set. Examples of such attributes include gene expression measurements, ages, and disease severity levels. By contrast, the second type consists of attributes that take values from an unordered set. Examples of such attributes include various gender, blood type, and tissue type. We refer to the first type as *continuous* attributes and to the second type as *categorical* attributes. The primary difference between these two types of attributes is that in the case of continuous attributes, when there is a mismatch on the value taken by a particular attribute in two different objects, the difference of the two values is a meaningful measure of distance, whereas in categorical attributes, there is no easy way to assign a distance between such mismatches.

Next we present methods for computing the similarity assuming that all the attributes in the objects are either continuous or categorical. However, in most real applications, objects will be represented by a mixture of such attributes, so the described approaches need to be combined.

3.1.1. Continuous Attributes

When all the attributes are continuous, each object can be considered to be a vector in the attribute space. That is, if n is the total number of attributes, then each object v can be represented by an n-dimensional vector $(v_1, v_2, \ldots, v_n)$, in which v_i is the value of the ith attribute.

Given any two objects with their corresponding vector-space representations **v** and **u**, a widely used method for computing the similarity between them is to look at their distance as measured by some norm of their vector difference. That is,

$$\text{dis}_r(\mathbf{v}, \mathbf{u}) = \|\mathbf{v} - \mathbf{u}\|_r \qquad (1)$$

in which r is the norm used, and $\|\cdot\|$ is used to denote vector norms. If the distance is small, then the objects will be similar, and the similarity of the objects will decrease as their distance increases.

The two most commonly used norms are the one- and the two-norm. In the case of the one-norm, the distance between two objects is given by

$$\text{dis}_1(\mathbf{v}, \mathbf{u}) = \|\mathbf{v} - \mathbf{u}\|_1 = \sum_i^n |v_i - u_i| \qquad (2)$$

in which $|\cdot|$ denotes absolute values. Similarly, in the case of the two-norm, the distance is given by

$$\text{dis}_2(\mathbf{v}, \mathbf{u}) = \|\mathbf{v} - \mathbf{u}\|_2 = \sqrt{\sum_{i=1}^n (v_i - u_i)^2} \qquad (3)$$

Note that the one-norm distance is also called the *Manhattan* distance, whereas the two-norm distance is nothing more than the *Euclidean* distance between the vectors. Those distances may become problematic when clustering high-dimensional data, because in such data sets, the similarity between two objects is often defined along a small subset of dimensions.

An alternate way of measuring the similarity between two objects in the vector-space model is to look at the angle between their vectors. If two objects have vectors that point to the same direction (i.e., their angle is small), then these objects will be considered similar, and if their vectors point to different directions (i.e., their angle is large), then these vectors will be considered dissimilar. This angle-based approach for computing the similarity between two objects emphasizes the relative values that each dimension takes within each vector, and not their overall length. That is, two objects can have an angle of 0 (i.e., point to the identical direction), even if their Euclidean distance is arbitrarily large. For example, in a 2D space, the vectors **v** = (1, 1), and **u** = (1000, 1000) will be considered to be identical, since their angle is 0. However, their Euclidean distance is close to $1000\sqrt{2}$.

Since the computation of the angle between two vectors is somewhat expensive (requiring inverse trigonometric functions), we do not measure the angle itself but its cosine function. The cosine of the angle between two vectors **v** and **u** is given by

$$\text{sim}(\mathbf{v}, \mathbf{u}) = \cos(\mathbf{v}, \mathbf{u}) = \frac{\sum_{i=1}^{n} v_i u_i}{\|v\|_2 \|u\|_2} \qquad (4)$$

This measure will be +1, if the angle between **v** and **u** is 0, and −1, if their angle is 180° (i.e., point to opposite directions). Note that a surrogate for the angle between two vectors can also be computed using the Euclidean distance, but instead of computing the distance between **v** and **u** directly, we need to first scale them to be of unit length. In that case, the Euclidean distance measures the *chord* between the two vectors in the unit hypersphere.

In addition to the discussed linear algebra–inspired methods, another widely used scheme for determining the similarity between two vectors uses the Pearson correlation coefficient, which is given by

$$\text{sim}(\mathbf{v}, \mathbf{u}) = \text{corr}(\mathbf{v}, \mathbf{u}) = \frac{\sum_{i=1}^{n}(v_i - \bar{v})(u_i - \bar{u})}{\sqrt{\sum_{i=1}^{n}(v_i - \bar{v})^2} \sqrt{\sum_{i=1}^{n}(u_i - \bar{u})^2}} \qquad (5)$$

in which $\bar{v}$ and $\bar{u}$ are the mean of the values of the v and u vectors, respectively. Note that the Pearson correlation coefficient is nothing more than the cosine between the mean-subtracted **v** and **u** vectors. As a result, it does not depend on the length of the ($\mathbf{v} - \bar{\mathbf{v}}$) and ($\mathbf{u} - \bar{\mathbf{u}}$) vectors, but only on their angle.

Our discussion so far on similarity measures for continuous attributes has focused on objects in which all the attributes were homogeneous in nature. A set of attributes is called *homogeneous* if all of the attributes measure quantities that are of the same type. As a result, changes in the values of these variables can be easily correlated across them. Quite often, each object will be represented by a set of inhomogeneous attributes. For example, if we would like to cluster patients, then some of the attributes describing each patient can measure things such as age, weight, height, and calorie intake. Now, if we use some of the described methods to compute the similarity, we will essentially be making the assumption that equal-magnitude changes in all variables are identical. However, this may not be the case. If the age of two patients is 50 yr, that represents something that is significantly different if their calorie intake difference is 50 calories. To address such problems, the various attributes need to be first normalized prior to using any of the stated similarity measures. Of course, the specific normalization method is attribute dependent, but its goal should be to make differences across different attributes comparable.

3.1.2. Categorical Attributes

If the attributes are categorical, special similarity measures are required, since distances between their values cannot be defined in an obvious manner. The most straightforward way is to treat each categorical attribute individually and define the similarity based on whether two objects contain the exact same value for each categorical attribute. Huang (**45**) formalized this idea by introducing dissimilarity measures between objects with categorical attributes that can be used in any clustering algorithms. Let X and Y be two objects with m categorical attributes, and X_i and Y_i be the values of the ith attribute of the two objects. The dissimilarity measure between X and Y is defined as the number of mismatching attributes of the two objects. That is,

$$d(X, Y) = \sum_{i=1}^{m} S(X_i, Y_i)$$

in which

$$S(X_i, Y_i) = \begin{cases} 0 & (X_i = Y_i) \\ 1 & (X_i \neq Y_i) \end{cases}$$

A normalized variant of this dissimilarity is defined as follows:

$$d(X,Y) = \sum_{i=1}^{m} \frac{n_{X_i} + n_{Y_i}}{n_{X_i} n_{Y_i}} s(X_i, Y_i)$$

in which n_{X_i} (n_{Y_i}) is the number of times the value X_i (Y_i) appears in the ith attribute of the entire data set. If two categorical values are common across the data set, they will have low weights, so that the mismatch between them will not contribute significantly to the final dissimilarity score. If two categorical values are rare in the data set, then they are more informative and will receive higher weights according to the formula. Hence, this dissimilarity measure emphasizes the mismatches that happen for rare categorical values rather than for those involving common ones.

One of the limitations of the preceding method is that two values can contribute to the overall similarity only if they are the same. However, different categorical values may contain useful information in the sense that even if their values are different, the objects containing those values are related to some extent. By defining similarities just based on matches and mismatches of values, some useful information may be lost. A number of approaches have been proposed to overcome this limitation *(19,46–48)* by utilizing additional information between categories or relationships between categorical values.

3.2. Similarity Between Sequences

One of the most important applications of clustering in life sciences is clustering sequences, e.g., DNA or protein sequences. Many clustering algorithms have been proposed to enhance sequence database searching, organize sequence databases, generate phylogenetic trees, guide multiple sequence alignment, and so on. In this specific clustering problem, the objects of interest are biological sequences, which consist of a sequence of *symbols*, which could be nucleotides, amino acids, or secondary structure elements (SSEs). Biological sequences are different from the objects we have discussed so far, in the sense that they are not defined by a collection of attributes. Hence, the similarity measures we discussed so far are not applicable to biological sequences.

Over the years, a number of different approaches have been developed for computing similarity between two sequences *(49)*. The most common are the alignment-based measures, which first compute an optimal alignment between two sequences (either globally or locally), and then determine their similarity by measuring the degree of agreement in the aligned positions of the two sequences. The aligned positions are usually scored using a symbol-to-symbol scoring matrix, and in the case of protein sequences, the most commonly used scoring matrices are PAM *(50,51)* and BLOSUM *(52)*.

The global sequence alignment (Needleman–Wunsch alignment *(53)*) aligns the entire sequences using dynamic programming. The recurrence relations are the following *(49)*. Given two sequences S_1 of length n and S_2 of length m, and a scoring matrix S, let score(i, j) be the score of the optimal alignment of prefixes $S_1[1 \ldots i]$ and $S_2[1 \ldots j]$.

The base conditions are

$$\text{score}(0, j) = \sum_{1 \leq k \leq j} S(\_, S_2(k))$$

and

$$\text{score}(i, 0) = \sum_{1 \leq k \leq i} S(S_1(k), \_)$$

Then, the general recurrence is

$$\text{score}(i, j) = \max \begin{cases} \text{score}(i-1, j-1) + S(S_1(i), S_2(i)) \\ \text{score}(i-1, j) + S(S_1(i), \_) \\ \text{score}(i, j-1) + S(\_, S_2(j)) \end{cases}$$

in which _ represents a space, and S is the scoring matrix to specify the matching score for each pair of symbols. And score(n, m) is the optimal alignment score.

These global similarity scores are meaningful when we compare similar sequences with roughly the same length, such as protein sequences from the same protein family. However, when sequences are of different lengths and are quite divergent, the alignment of the entire sequences may not make sense, in which case the similarity is commonly defined on the conserved subsequences. This problem is referred to as the *local alignment problem*, which seeks to find the pair of substrings of the two sequences that has the highest global alignment score among all possible pairs of substrings. Local alignments can be computed optimally via a dynamic programming algorithm, originally introduced by Smith and Waterman *(54)*. The base conditions are score$(0, j) = 0$ and score$(i, 0) = 0$, and the general recurrence is given by

$$\text{score}(i, j) = \max \begin{cases} 0 \\ \text{score}(i-1, j-1) + S(S_1(i), S_2(i)) \\ \text{score}(i-1, j) + S(S_1(i), \_) \\ \text{score}(i, j-1) + S(\_, S_2(j)) \end{cases}$$

The local sequence alignment score corresponds to the cell(s) of the dynamic programming table that has the highest value. Note that the recurrence for local alignments is very similar to that for global alignments only with minor changes, which allow the alignment to begin from any location (i, j) *(49)*.

Alternatively, local alignments can be computed approximately via heuristic approaches, such as FASTA *(55,56)* or BLAST *(4)*. The heuristic approaches achieve low time complexities by first identifying promising locations in an efficient way, and then applying a more expensive method on those locations to construct the final local sequence alignment. The heuristic approaches are widely used for searching protein databases due to their low time complexity. Description of these algorithms is beyond the scope of this chapter; interested readers should follow the references.

Most existing protein clustering algorithms use the similarity measure based on the local alignment methods, i.e., Smith–Waterman, BLAST, and FASTA (GeneRage *(2)*, ProtoMap *(58)*, and so on). These clustering algorithms first obtain the pairwise similarity scores of all pairs of sequences. Then they either normalize the scores by the self-similarity scores of the sequences to obtain a percentage value of identicalness *(59)*, or transform the scores to binary values based on a particular threshold *(2)*. Other methods normalize the row similarity scores by taking into account other sequences in the data set. For example, ProtoMage *(58)* first generates the distribution of the pairwise similarities between sequence A and the other sequences in the database. Then the similarity between sequence A and sequence B is defined as the expected value of the similarity score found for A and B, based on the overall distribution. A low expected value indicates a significant and strong connection (similarity).

3.3. Similarity Between Structures

Methods for computing the similarity between the 3D structures of two proteins (or other molecules) are intrinsically different from any of the approaches that we have seen so far for comparing multidimensional objects and sequences. Moreover, unlike the previous data types for which there are well-developed and widely accepted methods for measuring similarities, the methods for comparing 3D structures are still evolving, and the entire field is an active research area. Providing a comprehensive description of the various methods for computing the similarity between two structures requires a chapter (or a book) of its own and is far beyond the scope of this chapter. For this reason, our discussion in the rest of this section primarily focuses on presenting some of the issues involved in comparing 3D structures, in the context of proteins, and outlining some of the approaches that have been proposed for solving them. The reader should refer to Johnson and Lehtonen *(60)*, who provide an excellent introduction on the topic.

The general approach, which almost all methods for computing the similarity between a pair of 3D protein structures follows, is to try to *superimpose* the structure of one protein on top of the structure of the other protein, so that certain key features are mapped very close to each other in space. Once this is done, the similarity between two structures is computed by measuring the *fit* of the superposition. This fit is commonly computed as the *root mean square deviations* of the corresponding features. To some extent, this is similar in nature to the alignment performed for sequence-based similarity approaches, but it is significantly more complicated because it involves 3D structures with substantially more degrees of freedom. There are a number of different variations for performing this superposition that deal with the features of the two proteins that are sought to be matched, whether or not the proteins are treated as rigid or flexible bodies, how the equivalent set of features from the two proteins is determined, and the type of superposition that is computed.

In principle, when comparing two protein structures, we can treat every atom of each amino acid side chain as a feature and try to compute a superposition that matches all of them as well as possible. However, this usually does not lead to good results because the side chains of different residues will have a different number of atoms with different geometries. Moreover, even the same amino acid types may have side chains with different conformations, depending on their environment. As a result, even if two proteins have very similar backbones, a superposition computed by looking at all the atoms may fail to identify this similarity. For this reason, most approaches try to superimpose two protein structures by focusing on the C_α atoms of their backbones, whose locations are less sensitive on the actual residue type. Besides these atom-level approaches, other methods focus on SSEs and superimpose two proteins so that their SSEs are geometrically aligned with each other.

Most approaches for computing the similarity between two structures treat them as rigid bodies and try to find the appropriate geometric transformation (i.e., rotation and translation) that leads to the best superposition. Rigid-body geometric transformations are well understood and are relatively easy to compute efficiently. However, by treating proteins as rigid bodies, we may get poor superpositions when the protein structures are significantly different, even though they are part of the same fold. In such cases, allowing some degree of flexibility tends to produce better results but also increases the complexity. In trying to find the best way to superimpose one structure on top of the other in addition to the features of interest, we must identify the pairs of features from the two structures that will be mapped against each other. There are two general approaches for doing that. The first approach relies on an initial set of equivalent features (e.g., C_α atoms or SSEs) being provided by domain experts. This initial set is used to compute an initial superposition, and then additional features are identified using various approaches based on dynamic programming or graph theory *(53,61,62)*. The second approach tries to identify automatically the correspondence between the various features by various methods including structural comparisons based on matching C_α atoms contact maps *(63)*, or on the optimal alignment of SSEs *(64)*.

Finally, as was the case with sequence alignment, the superposition of 3D structures can be done globally, with the goal of superimposing the entire protein structure, or locally, with the goal of computing a good superposition involving a subsequence of the protein.

4. Assessing Cluster Quality

Clustering results are hard to evaluate, especially for high-dimensional data and without *a priori* knowledge of the objects' distribution, which is quite common in practical cases. However, assessing the quality of the resulting clusters is as important as generating the clusters. Given the same data set, different clustering algorithms with various parameters or initial conditions will give very different clusters. It is essential to know whether the resulting clusters are valid and how to compare the quality of the clustering results, so that the right clustering algorithm can be chosen and the best clustering results can be used for further analysis.

Another related problem is answering the question, How many clusters are there in the data set? An ideal clustering algorithm should be one that can automatically discover the natural clusters present in the data set based on the underlying cluster definition. However, there are no such universal cluster definitions and clustering algorithms suitable for all kinds of data sets. As a result, most existing algorithms require either the number of clusters to be provided as a parameter, as is done in the case of K-means, or a similarity threshold that will be used to terminate the merging process, as in the case of agglomerative clustering. However, in general, it is hard to know the right number of clusters or the right similarity threshold without *a priori* knowledge of the data set.

One possible way to determine automatically the number of clusters k is to compute various clustering solutions for a range of values of k, score the resulting clusters based on some particular metric, and then select the solution that achieves the best score. A critical component of this approach is the method used to measure the quality of the cluster. To solve this problem, numerous approaches have been proposed in a number of different disciplines including pattern recognition, statistics, and data mining. The majority can be classified into two groups: *external quality measures* and *internal quality measures*.

The approaches based on external quality measures require *a priori* knowledge of the natural clusters that exist in the data set and validate a clustering result by measuring the agreement between the discovered clusters and the known information. For instance, when clustering gene expression data, the known functional categorization of the genes can be treated as the natural clusters, and the resulting clustering solution will be considered correct, if it leads to clusters that preserve this categorization. A key aspect of the external quality measures is that they utilize information other than that used by the clustering algorithms. However, such reliable *a priori* knowledge is usually not available when analyzing real data sets—after all, clustering is used as a tool to discover such knowledge in the first place.

The basic idea behind internal quality measures is rooted from the definition of clusters. A meaningful clustering solution should group objects into various clusters, so that the objects within each cluster are more similar to each other than the objects from different clusters. Therefore, most of the internal quality measures evaluate the clustering solution by looking at how similar the objects are within each cluster and how well the objects of different clusters are separated. For example, the pseudo F statistic suggested by Calinski and Harabasz *(65)* uses the quotient between the intracluster average squared distance and intercluster average squared distance. If we have X as the centroid (i.e., mean vector) of all the objects, X_j as the centroid of the objects in cluster C_j, k as the total number of clusters, n as the total number of objects, and $d(x, y)$ as the squared Euclidean distance between two object vectors x and y, then the pseudo F statistic is defined as follows:

$$F = \frac{\sum_{i=1}^{n} d(x_i, X) - \sum_{j=1}^{k} \sum_{x \in C_j} d(x, X_j)}{\frac{k-1}{\sum_{j=1}^{k} \sum_{x \in C_j} d(x, X_j)}{n-k}}$$

One of the limitations of the internal quality measures is that they often use the same information both in discovering and in evaluating the clusters. Recall from **Subheading 2.** that some clustering algorithms produce clustering results by optimizing various clustering criterion functions. Now, if the same criterion functions were used as the internal quality measure, then the overall clustering assessment process would do nothing more than *assess* how effective the clustering algorithm was in optimizing the particular criterion function and would provide no independent confirmation about the degree to which the clusters are meaningful.

An alternative way of validating the clustering results is to see how stable they are when adding noise to the data, or subsampling it *(66)*. This approach performs a sequence of subsamplings of the data set and uses the same clustering procedure to produce clustering solutions for various subsamples. These various clustering results are then compared to determine the degree to which they agree. The stable clustering solution should be the one that gives similar clustering results across the different subsamples. This approach can also be easily used to determine the *correct* number of clusters in hierarchical clustering solutions. The stability test of clustering is performed at each level of the hierarchical tree, and the number of clusters k will be the largest k value that still can produce stable clustering results.

Finally, a recent approach, with applications to clustering gene expression data sets, assesses the clustering results of gene expression data by looking at the predictive power for one experimental condition from the clustering results based on the other experimental conditions *(67)*. The key idea behind this approach is that if one condition is left out, then the clusters generated from the remaining conditions should exhibit lower variation in the left-out condition than randomly formed clusters. Yeung et al. *(67)* defined the *figure of merit* to be the summation of intracluster variance for each one of the clustering instances in which one of the conditions was not used during clustering (i.e., left-out condition). Among the various clustering solutions, they prefer the one that exhibits the least variation, and their experiments showed that in the context of clustering gene expression data, this method works quite well. The limitation of this approach is that it is not applicable to a data set in which all the attributes are independent. Moreover, this approach is only applicable to low-dimensional data sets, since computing the intracluster variance for each dimension is quite expensive when the number of dimensions is very large.

5. Case Study: Clustering Gene Expression Data

Recently developed methods for monitoring genomewide mRNA expression changes such as oligonucleotide chips *(68)* and cDNA microarrays *(69)* are especially powerful because they allow us to monitor quickly and inexpensively the expression levels of a large number of genes at different time points for different conditions, tissues, and organisms. Knowing when and under what conditions a gene or a set of genes is expressed often provides strong clues to their biological role and function.

Clustering algorithms are used as an essential tool to analyze these data sets and provide valuable insight into various aspects of the genetic machinery. There are four distinct classes of clustering problems that can be formulated from the gene expression data sets, each addressing a different biological problem. The first problem focuses on finding *coregulated genes* by grouping genes that have similar expression profiles. These coregulated genes can be used to identify promoter elements by finding conserved areas in their upstream regions. The second problem focuses on finding *distinctive tissue types* by grouping tissues whose genes have similar expression profiles. These tissue groups can then be further analyzed to identify the genes that best distinguish the various tissues. The third clustering problem focuses on finding *common inducers* by grouping conditions for which the expression profiles of the genes are similar. Finding such groups of common inducers will allow us to substitute different "trigger" mechanisms that still elicit the same response (e.g., similar drugs, or similar herbicides or pesticides). Finally, the fourth clustering problem focuses on finding organisms that exhibit similar responses over a specified set of tested conditions by grouping organisms for which the expression profiles of their genes (in an ortholog sense) are similar. This would allow us to identify organisms with similar responses to chosen conditions (e.g., microbes that share a pathway).

Next we briefly review the approaches behind cDNA and oligonucleotide microarrays and discuss various issues related to clustering such gene expression data sets.

5.1. Overview of Microarray Technologies

DNA microarrays measure gene expression levels by exploiting the preferential binding of complementary, single-stranded nucleic acid sequences. cDNA microarrays, developed at Stanford University, are glass slides, to which single-stranded DNA (ssDNA) molecules are attached at fixed locations (spots) by high-speed robotic printing *(70)*. Each array may contain tens of thousands of spots, each of which corresponds to a single gene. mRNA from the sample and from control cells is extracted and cDNA is prepared by reverse transcription. Then, cDNA is labeled with two fluorescent dyes and washed over the microarray so that cDNA sequences from both populations hybridize to their complementary sequences in the spots. The amount of cDNA from both populations bound to a spot can be measured by the level of fluorescence emitted from each dye. For example, the sample cDNA is labeled with a red dye and the control cDNA is labeled with a green dye. Then, if the mRNA from the sample population is in abundance, the spot will be red; if the mRNA from the control population is in abundance, it will be green; if sample and control bind equally, the spot will be yellow; if neither binds, it will appear black. Thus, the relative expression levels of the genes in the sample and control populations can be estimated from the fluorescent intensities and colors for each spot. After transforming the raw images produced by microarrays into relative fluorescent intensity via some image-processing software, the gene expression levels are estimated as log ratios of the relative intensities. A gene expression matrix can be formed by combining multiple microarray experiments of the same set of genes but under different conditions, at which each row corresponds to a gene and each column corresponds to a condition (i.e., a microarray experiment) *(70)*.

The Affymetrix GeneChip oligonucleotide array contains several thousand ssDNA oligonucleotide probe pairs. Each probe pair consists of an element containing oligonucleotides that perfectly match the target (PM probe) and an element containing oligonucleotides with a single base mismatch (MM probe). A probe set consists of a set of probe pairs corresponding to a target gene. Similarly, the labeled RNA is extracted from sample cell and hybridizes to its complementary sequence. The expression level is measured by determining the difference between the PM and MM probes. Then, for each gene (i.e., probe set) average difference or log average can be calculated, in which the average difference is defined as the average difference between the PM and MM of every probe pair in a probe set and log average is defined as the average log ratios of the PM/MM intensities for each probe pair in a probe set.

5.2. Preparation and Normalization of Data

Many sources of systematic variation may affect the measured gene expression levels in microarray experiments *(71)*. For the GeneChip experiments, scaling/normalization must be performed for each experiment before combining them, so that they can have the same target intensity. The scaling factor of each experiment is determined by the array intensity of the experiment and the common target intensity, in which the array intensity is a composite of the average difference intensities across the entire array.

For cDNA microarray experiments, two fluorescent dyes are involved and cause more systematic variation, which makes normalization more important. In particular, this variation could be caused by differences in RNA amounts, differences in labeling efficiency between the two fluorescent dyes, and image acquisition parameters. Such biases can be removed by a constant adjustment to each experiment to force the distribution of the log ratios to have a median of zero. Since an experiment corresponds to one column in the gene expression array, this global normalization can be done by subtracting the mean/median of the gene expression levels of one experiment from the original values, so that the mean value for this experiment (column) is 0.

However, there are other sources of systematic variation that global normalization may not be able to correct. Yang et al. *(71)* pointed out that dye biases can depend on a spot's overall intensity and location on the array. Given the red and green fluorescence intensities (R, G) of all the spots in one slide, they plotted the log intensity ratio $M = \log R/G$ vs the mean log intensity $A = \log \sqrt{RG}$, which shows clear dependence of the log ratio M on overall spot intensity A. Hence, an intensity-related normalization was proposed, in which the original log ratio $\log R/G$ is subtracted by $C(A)$. $C(A)$ is a scatter plot smoother fit to the M vs A plot using robust locally linear fits.

5.3. Similarity Measures

In most microarray clustering applications our goal is to find clusters of genes and/or clusters of conditions. Several different methods have been proposed for computing these similarities, including Euclidean distance-based similarities, correlation coefficients, and mutual information.

The use of correlation coefficient–based similarities is primarily motivated by the fact that while clustering gene expression data sets, we are interested on how the expression levels of different genes are related under various conditions. The correlation coefficient values between genes (Eq. 5) can be used directly or transformed to absolute values if genes of both positive and negative correlations are important in the application.

An alternate way of measuring the similarity is to use the mutual information between a pair of genes. The mutual information between two information sources A and B represents how much information the two sources contain for each other. D'haeseleer et al. *(72)* used mutual information to define the relationship between two conditions A and B. This was done by initially discretizing the gene expression levels into various bins, and using this discretization to compute the Shannon entropy of conditions A and B as follows:

$$S_A = -\sum_i p_i \log p_i$$

in which p_i is the frequency of each bin. Given these entropy values, the mutual information between A and B is defined as

$$M(A, B) = S_A + S_B - S_{A,B}$$

A feature common to many similarity measures used for microarray data is that they almost never consider the length of the corresponding gene or condition vectors, (i.e., the actual value of the differential expression level), but focus only on various measures of relative change and/or how these relative measures are correlated between two genes or conditions *(67,73,74)*. The reason for this is twofold. First, there are still significant experimental errors in measuring the expression level of a gene, and it is not reliable to use it "as is." Second, in most cases we are only interested in how the different genes change across the different conditions (i.e., either up- or downregulated), and we are not interested in the exact amount of this change.

5.4. Clustering Approaches for Gene Expression Data

Since the early days of the development of microarray technologies, a wide range of existing clustering algorithms have been used, and novel new approaches have been developed for clustering gene expression data sets. The most effective traditional clustering algorithms are based either on the group-average variation of the agglomerative clustering methodology, or on the K-means approach applied to unit-length gene or condition expression vectors. Unlike other applications of clustering in life sciences, such as the construction of phylogenetic trees, or guide trees for multiple sequence alignment, there is no biological reason that justifies that the structure of the correct clustering solution is in the form of a tree. Thus, agglomerative solutions are inherently suboptimal when compared to partitional approaches, which allow for a wider range of feasible solutions at various levels of cluster granularity. However, the agglomerative solutions do tend to produce reasonable and biologically meaningful results and allow easy visualization of the relationships between the various genes and/or conditions in the experiments.

The ease of visualizing the results has also led to the extensive use of self-organizing maps (SOMs) for gene expression clustering *(74,51)*. The SOM method starts with a geometry of "nodes" of a simple topology (e.g., grid and ring) and a distance function on the nodes. Initially, the nodes are mapped randomly into the gene expression space, in which the ith coordinate represents the expression level in the ith condition. At each following iteration, a data point P, i.e., a gene expression profile, is randomly selected and the data point P will attract nodes to itself. The nearest node N_p to P will be identified and moved the most, and other nodes will be adjusted depending on their distances to the nearest node N_p toward the data point P. The advantages of using SOMs are their structured approach, which makes visualization very easy. However, SOMs require the user to specify the number of clusters as well as the grid topology, including the dimensions of the grid and the number of clusters in each dimension.

From the successes obtained in using K means and group-average-based clustering algorithms, as well as other similar algorithms *(76,77)*, it appears that the clusters in the context of gene expression data sets are globular in nature. This should not be surprising since researchers are often interested in obtaining clusters whose genes have similar expression patterns/profiles. Such a requirement automatically lends itself to globular clusters, in which the pairwise similarity between most object pairs is quite high. However, as the dimensionality of these data sets continues to increase (primarily by increasing the number of conditions that are analyzed), requiring consistency across the entire set of conditions will be unrealistic. As a result, approaches that try to find tight clusters in subspaces of these conditions may gain popularity.

6. CLUTO: A Clustering Toolkit

We now turn our focus on providing a brief overview of CLUTO (release 2.0), a software package for clustering low- and high-dimensional data sets and for analyzing the characteristics of the various clusters, that was developed by our group and is available at www.cs.umn.edu/~karypis/cluto. CLUTO was developed as a general purpose clustering toolkit. CLUTO's distribution consists of stand-alone programs (vcluster and cluster) for clustering and analyzing these clusters, as well as a library via which an application program can access directly the various clustering and analysis algorithms implemented in CLUTO. To date, CLUTO has been successfully used to cluster data sets arising in many diverse application areas including information retrieval, commercial data sets, scientific data sets, and biological applications.

CLUTO implements three different classes of clustering algorithms that can operate either directly in the object's feature space or in the object's similarity space. The clustering algorithms provided by CLUTO are based on the partitional, agglomerative, and graph-partitioning paradigms. CLUTO's partitional and agglomerative algorithms are able to find clusters that are primarily globular, whereas its graph-partitioning and some of its agglomerative algorithms are capable of finding transitive clusters.

A key feature in most of CLUTO's clustering algorithms is that they treat the clustering problem as an optimization process that seeks to maximize or minimize a particular *clustering criterion function* defined either globally or locally over the entire clustering solution space. CLUTO provides a total of seven

Table 1
Mathematical Definition of CLUTO's Clustering Criterion Functions[a]

Criterion function	Optimization function	
I_1	$\text{maximize} \sum_{i=1}^{k} \frac{1}{n_i} \left(\sum_{v,u \in S_i} \text{sim}(v,u) \right)$	(6)
I_2	$\text{maximize} \sum_{i=1}^{k} \sqrt{\sum_{v,u \in S_i} \text{sim}(v,u)}$	(7)
$\mathcal{E}_1$	$\text{minimize} \sum_{i=1}^{k} n_i \frac{\sum_{v \in S_i, u \in S} \text{sim}(v,u)}{\sqrt{\sum_{v,u \in S_i} \text{sim}(v,u)}}$	(8)
$\mathcal{G}_1$	$\text{minimize} \sum_{i=1}^{k} \frac{\sum_{v \in S_i, u \in S} \text{sim}(v,u)}{\sum_{v,u \in S_i} \text{sim}(v,u)}$	(9)
$\mathcal{G}'_1$	$\text{minimize} \sum_{i=1}^{k} n_i^2 \frac{\sum_{v \in S_i, u \in S} \text{sim}(v,u)}{\sum_{v,u \in S_i} \text{sim}(v,u)}$	(10)
$\mathcal{H}_1$	$\text{maximize} \ \frac{I_1}{\mathcal{E}_1}$	(11)
$\mathcal{H}_2$	$\text{maximize} \ \frac{I_2}{\mathcal{E}_1}$	(12)

[a]The notation in these equations is as follows: k is the total number of clusters, S is the total objects to be clustered, S_i is the set of objects assigned to the ith cluster, n_i is the number of objects in the ith cluster, v and u represent two objects, and $\text{sim}(v,u)$ is the similarity between two objects.

different criterion functions that have been shown to produce high-quality clusters in low- and high-dimensional data sets. The equations of these criterion functions are shown in **Table 1**, and they were derived and analyzed in (**refs. 30** and **78**). In addition to these criterion functions, CLUTO provides some of the more traditional local criteria (e.g., single link, complete link, and group average) that can be used in the context of agglomerative clustering.

An important aspect of partitional-based criterion-driven clustering algorithms is the method used to optimize this criterion function. CLUTO uses a randomized incremental optimization algorithm that is greedy in nature, has low computational requirements, and produces high-quality clustering solutions (*30*). Moreover, CLUTO's graph-partitioning-based clustering algorithms utilize high-quality and efficient multilevel graph-partitioning algorithms derived from the METIS and hMETIS graph and hypergraph partitioning algorithms (*79,80*). Moreover, CLUTO's algorithms have been optimized for operating on very large data sets both in terms of the number of objects and in the number of dimensions. This is especially true for CLUTO's algorithms for partitional clustering. These algorithms can quickly cluster data sets with several tens of thousands of objects and several thousands of dimensions. Moreover, since most high-dimensional data sets are very sparse, CLUTO directly takes into account this sparsity and requires memory that is roughly linear on the input size.

In the rest of this section, we present a short description of CLUTO's stand-alone programs followed by some illustrative examples of how it can be used for clustering biological datasets.

6.1. Usage Overview

The vcluster and scluster programs are used to cluster a collection of objects into a predetermined number of clusters k. The vcluster program treats each object as a vector in a high-dimensional space, and it computes the clustering solution using one of five different approaches. Four of these approaches are *partitional* in nature, whereas the fifth approach is *agglomerative*. On the other hand, the scluster program operates on the similarity space between the objects but can compute the overall clustering solution using the same set of five different approaches.

Both the vcluster and scluster programs are invoked by providing two required parameters on the command line along with a number of optional parameters. Their overall calling sequence is as follows:

    vcluster   [optional parameters]   *MatrixFile*   *Nclusters*
    scluster   [optional parameters]   *GraphFile*   *Nclusters*

MatrixFile is the name of the file that stores the n objects that need to be clustered. In vcluster, each of these objects is considered to be a vector in an m-dimensional space. The collection of these objects is treated as an $n \times m$ matrix, whose rows correspond to the objects, and whose columns correspond to the dimensions of the feature space. Similarly, *GraphFile* is the name of the file that stores the adjacency matrix of the similarity graph between the n objects to be clustered. The second argument for both programs, *Nclusters*, is the number of clusters that is desired.

Figure 1 shows the output of vcluster for clustering a matrix into 10 clusters. We see that vcluster initially prints information about the matrix, such as its name, the number of rows (*#Rows*), the number of columns (*#Columns*), and the number of nonzeros in the matrix (*#NonZeros*). Next, it prints information about the values of the various options that it used to compute the clustering, and the number of desired clusters (*#Clusters*). Once it computes the clustering solution, it displays information regarding the quality of the overall clustering solution, as well as the quality of each cluster, using a variety of internal quality

```
prompt% vcluster sports.mat 10
******************************************************************************
vcluster (CLUTO 2.0) Copyright 2001-02, Regents of the University of Minnesota

Matrix Information ----------------------------------------------------------
  Name: sports.mat, #Rows: 8580, #Columns: 126373, #NonZeros: 1107980

Options ---------------------------------------------------------------------
  CLMethod=RB, CRfun=I2, SimFun=Cosine, #Clusters: 10
  RowModel=None, ColModel=IDF, GrModel=SY-DIR, NNbrs=40
  Colprune=1.00, EdgePrune=-1.00, VtxPrune=-1.00, MinComponent=5
  CSType=Best, AggloFrom=0, AggloCRFun=I2, NTrials=10, NIter=10

Solution --------------------------------------------------------------------

-----------------------------------------------------------------------------
10-way clustering: [I2=2.29e+03] [8580 of 8580]
-----------------------------------------------------------------------------
cid   Size   ISim   ISdev    ESim   ESdev   |
-----------------------------------------------------------------------------
  0    364  +0.166 +0.050  +0.020 +0.005  |
  1    628  +0.106 +0.041  +0.022 +0.007  |
  2    793  +0.102 +0.036  +0.018 +0.006  |
  3    754  +0.100 +0.034  +0.021 +0.006  |
  4    845  +0.095 +0.035  +0.023 +0.007  |
  5    637  +0.079 +0.036  +0.022 +0.008  |
  6   1724  +0.059 +0.026  +0.022 +0.007  |
  7    703  +0.049 +0.018  +0.016 +0.006  |
  8   1025  +0.054 +0.016  +0.021 +0.006  |
  9   1107  +0.029 +0.010  +0.017 +0.006  |
-----------------------------------------------------------------------------

Timing Information ----------------------------------------------------------
  I/O:                                   0.920 sec
  Clustering:                           12.440 sec
  Reporting:                             0.220 sec
******************************************************************************
```

Fig. 1. Output of vcluster for matrix sports.mat and a 10-way clustering.

measures. These measures include the average pairwise similarity between each object of the cluster and its standard deviation (*ISim* and *ISdev*), and the average similarity between the objects of each cluster to the objects in the other clusters and their standard deviation (*ESim* and *ESdev*). Finally, vcluster reports the time taken by the various phases of the program.

6.2. Summary of Biologically Relevant Features

The behavior of vcluster and scluster can be controlled by specifying more than 30 different optional parameters. These parameters can be broadly categorized into three groups. The first group controls various aspects of the clustering algorithm, the second group controls the type of analysis and reporting that is performed on the computed clusters, and the third group controls the visualization of the clusters. Some of the most important parameters are

Table 2
Key Parameters of CLUTO's Clustering Algorithms

Parameter	Values	Function
-clmethod	rb, direct, agglo, graph	Clustering method
-sim	cos, corr, dist	Similarity measures
-crfun	$I_1, I_2, E_1, G_1, G'_1, H_1, H_2$, slink, wslink, clink, wclink, upgma	Criterion function
-agglofrom	(int)	Where to start agglomeration
-fulltree		Builds a tree within each cluster
-showfeatures		Displays cluster's feature signature
-showtree		Builds a tree on top of clusters
-labeltree		Provides key features for each tree node
-plottree	(filename)	Plots agglomerative tree
-plotmatrix	(filename)	Plots input matrices
-plotclusters	(filename)	Plots cluster-cluster matrix
-clustercolumn		Simultaneously clusters the features

shown in **Table 2** and are described in the context of clustering biological data sets in the rest of the chapter.

6.3. Clustering Algorithms

The -clmethod parameter controls the type of algorithms to be used for clustering. The first two methods, ("rb" and "direct") follow the partitional paradigm described in **Subheading 2.1**. The difference between them is the method that they use to compute the k-way clustering solution. In the case of "rb," the k-way clustering solution is computed via a sequence of repeated bisections, whereas in the case of "direct," the entire k-way clustering solution is computed at one step. CLUTO's traditional agglomerative algorithm is implemented by the "agglo" option, whereas the "graph" option implements a graph-partitioning-based clustering algorithm that is well suited for finding transitive clusters. The method used to define the similarity between the objects is specified by the -sim parameter and supports the cosine ("cos"), correlation coefficient ("corr"), and a Euclidean distance-derived similarity ("dist"). The clustering criterion function that is used by the partitional and agglomerative algorithms is controlled by the -crfun parameter. The first seven criterion functions (described in **Table 1**) are used by both partitional and agglomerative, whereas the last five (single link, weighted single link, complete link, weighted complete link, and group average) are only applicable to agglomerative clustering.

A key feature of CLUTO's is that it allows you to combine partitional and agglomerative clustering approaches. This is done by the -agglofrom parameter in the following way. The desired k-way clustering solution is computed by first clustering the data set into m clusters ($m > k$) and then uses an agglomerative algorithm to group some of these clusters to form the final k-way clustering solution. The number of clusters m is the value supplied to -agglofrom. This approach was motivated by the two-phase clustering approach of the CHAMELEON algorithm (20) and was designed to allow the user to compute a clustering solution that uses a different clustering criterion function for the partitioning phase from that used for the agglomeration phase. An application of such an approach is to allow the clustering algorithm to find nonglobular clusters. In this case, the partitional clustering solution can be computed using a criterion function that favors globular clusters (e.g., "i2"), and then combine these clusters using a single-link approach (e.g., "wslink") to find nonglobular but well-connected clusters.

6.4. Building Tree for Large Data Sets

Hierarchical agglomerative trees are used extensively in life sciences because they provide an intuitive way to organize and visualize the clustering results. However, there are two limitations with such trees. First, hierarchical agglomerative clustering may not be the *optimal* way to cluster data in which there is no biological reason to suggest that the objects are related to each other in a tree fashion. Second, hierarchical agglomerative clustering algorithms have high computational and memory requirements, making them impractical for data sets with more than a few thousand objects.

To address these problems CLUTO provides the -fulltree option that can be used to produce a complete tree using a hybrid of partitional and agglomerative approaches. In particular, when -fulltree is specified, CLUTO builds a complete hierarchical tree that preserves the clustering solution that was computed. In this hierarchical clustering solution, the objects of each cluster form a subtree, and the different subtrees are merged to get an all-inclusive cluster at the end. Furthermore, the individual trees are combined in a meaningful way, so that the similarities within each tree are accurately represented.

Figure 2 shows the trees produced on a sample gene expression data set. The first tree (**Fig. 2A**) was obtained using the agglomerative clustering algorithm, whereas the second tree (**Fig. 2B**) was obtained using the repeated-bisecting method in which the -fulltree was specified.

6.5. Analyzing the Clusters

In addition to the core clustering algorithms, CLUTO provides tools to analyze each of the clusters and identify the features that best describe and discriminate

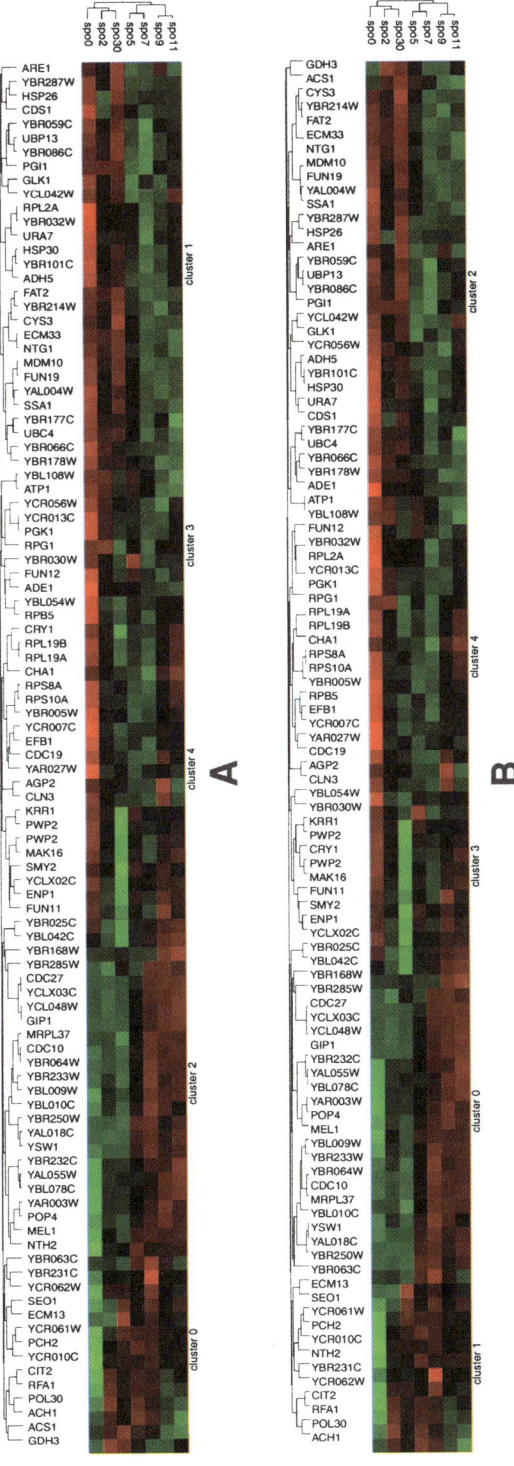

Fig. 2. (**A**) Clustering solution produced by agglomerative method; (**B**) clustering solution produced by repeated-bisecting method and -fulltree.

```
*******************************************************************
vcluster (CLUTO 2.0) Copyright 2001-02, Regents of the University of Minnesota

Matrix Information ------------------------------------------------
 Name: peptide5.mat, #Rows: 1539, #Columns: 2965, #NonZeros: 50136

Options -----------------------------------------------------------
  CLMethod=RB, CRfun=I2, SimFun=Cosine, #Clusters: 10
  RowModel=None, ColModel=IDF, GrModel=SY-DIR, NNbrs=40
  Colprune=1.00, EdgePrune=-1.00, VtxPrune=-1.00, MinComponent=5
  CSType=Best, AggloFrom=0, AggloCRFun=I2, NTrials=10, NIter=10

Solution ----------------------------------------------------------

-------------------------------------------------------------------
10-way clustering: [I2=5.03e+02] [1539 of 1539], Entropy: 0.510, Purity: 0.638

 cid  Size   ISim   ISdev   ESim   ESdev  Entpy Purty |  1    2    3    4    5
-------------------------------------------------------------------
  0    50  +0.439 +0.139  +0.001 +0.001  0.061 0.980 |  0    0    1   49    0
  1   109  +0.275 +0.113  +0.019 +0.010  0.156 0.945 |  0    3    3  103    0
  2    99  +0.246 +0.089  +0.001 +0.001  0.096 0.970 |  0    1   96    2    0
  3   227  +0.217 +0.074  +0.010 +0.005  0.129 0.956 |  0    3    7  217    0
  4   121  +0.197 +0.092  +0.001 +0.001  0.258 0.893 |  0  108    6    7    0
  5   127  +0.113 +0.064  +0.001 +0.001  0.425 0.780 | 99    5   21    2    0
  6   112  +0.109 +0.101  +0.001 +0.001  0.553 0.661 |  1   74   12   25    0
  7   202  +0.043 +0.040  +0.001 +0.002  0.741 0.421 |  4    6   78   85   29
  8   268  +0.029 +0.029  +0.001 +0.001  0.880 0.302 | 81   47   81   55    4
  9   224  +0.027 +0.024  +0.001 +0.001  0.860 0.312 |  1   57   70   37   59
-------------------------------------------------------------------

10-way clustering solution - Descriptive & Discriminating Features...
-------------------------------------------------------------------
Cluster   0, Size:    50, ISim: 0.439, ESim: 0.001
    Descriptive:  GTSMA 58.5%, HGTHV 37.7%, PSTVV  0.4%, LGASG 0.4%, KELKK 0.3%
 Discriminating:  GTSMA 29.6%, HGTHV 19.1%, GDSGG  2.7%, DSGGP 2.5%, TAAHC 2.0%

Cluster   1, Size:   109, ISim: 0.275, ESim: 0.019
    Descriptive:  SGGPL  8.6%, RPYMA  7.6%, VLTAA  7.2%, TAAHC 6.9%, DSGGP 6.3%
 Discriminating:  RPYMA  5.5%, PHSRP  4.3%, SRPYM  4.3%, HSRPY 3.9%, KGDSG 3.3%

Cluster   2, Size:    99, ISim: 0.246, ESim: 0.001
    Descriptive:  HEFGH 13.2%, CGVPD  6.5%, RCGVP  6.3%, PRCGV 6.1%, VAAHE 5.3%
 Discriminating:  HEFGH  6.8%, CGVPD  3.3%, RCGVP  3.2%, PRCGV 3.1%, GDSGG 2.9%

Cluster   3, Size:   227, ISim: 0.217, ESim: 0.010
    Descriptive:  GDSGG  8.3%, DSGGP  6.7%, CQGDS  5.7%, QGDSG 5.6%, TAAHC 4.4%
 Discriminating:  CQGDS  3.5%, QGDSG  3.3%, NSPGG  2.6%, GDSGG 2.3%, CGGSL 2.1%

Cluster   4, Size:   121, ISim: 0.197, ESim: 0.001
    Descriptive:  CGSCW 13.9%, GSCWA 10.2%, SCWAF  8.4%, KNSWG 5.3%, GCNGG 5.0%
 Discriminating:  CGSCW  7.1%, GSCWA  5.2%, SCWAF  4.3%, GDSGG 2.9%, KNSWG 2.7%

Cluster   5, Size:   127, ISim: 0.113, ESim: 0.001
    Descriptive:  DTGSS  6.0%, FDTGS  4.0%, TGSSD  3.0%, IGTPP 2.7%, GTPPQ 2.6%
 Discriminating:  DTGSS  3.1%, GDSGG  2.8%, DSGGP  2.6%, TAAHC 2.0%, SGGPL 2.0%

Cluster   6, Size:   112, ISim: 0.109, ESim: 0.001
    Descriptive:  KDELR  2.2%, IEASS  1.6%, RWAVL  1.6%, TFLKR 1.4%, EEKIK 1.3%
 Discriminating:  GDSGG  2.8%, DSGGP  2.6%, TAAHC  2.0%, SGGPL 2.0%, LTAAH 1.4%

Cluster   7, Size:   202, ISim: 0.043, ESim: 0.001
    Descriptive:  NSPGG 46.9%, HELGH 12.1%, ALLEV  7.0%, VLAAA 2.6%, GYVDA 1.7%
 Discriminating:  NSPGG 24.5%, HELGH  5.1%, ALLEV  3.6%, GDSGG 2.9%, DSGGP 2.6%

Cluster   8, Size:   268, ISim: 0.029, ESim: 0.001
    Descriptive:  QACRG 13.7%, IQACR  7.5%, DTGAD  2.9%, VDTGA 2.5%, LDTGA 1.2%
 Discriminating:  QACRG  7.1%, IQACR  3.9%, GDSGG  2.9%, DSGGP 2.6%, TAAHC 2.1%

Cluster   9, Size:   224, ISim: 0.027, ESim: 0.001
    Descriptive:  LAAIA  4.3%, TDNGA  3.0%, LKTAV  1.4%, TQYGG 1.1%, GFRRL 1.1%
 Discriminating:  GDSGG  2.9%, DSGGP  2.6%, LAAIA  2.2%, TAAHC 2.1%, SGGPL 2.0%
-------------------------------------------------------------------

Timing Information ------------------------------------------------
  I/O:                                       0.040 sec
  Clustering:                                0.470 sec
  Reporting:                                 0.040 sec
*******************************************************************
```

Fig. 3. Output of vcluster for matrix sports.mat and a 10-way clustering that shows descriptive and discriminating features of each cluster.

Clustering in Life Sciences

each one of the clusters. To some extent, these analysis methods try to identify the dense subspaces in which each cluster is formed. This is accomplished by the -showfeatures and -labeltree parameters.

Figure 3 shows the output produced by vcluster when -showfeatures was specified for a data set consisting of protein sequences and the 5mers that they contain. We can see that the set of descriptive and discriminating features are displayed right after the table that provides statistics for the various clusters. For each cluster, vcluster displays three lines of information. The first line contains some basic statistics for each cluster corresponding to the cluster-id (cid), number of objects in each cluster (Size), average pairwise similarity between the cluster's objects (ISim), and average pairwise similarity to the rest of the objects (ESim). The second line contains the five most descriptive features, whereas the third line contains the five most discriminating features. The features in these lists are sorted in decreasing descriptive or discriminating order.

Right next to each feature, vcluster displays a number that in the case of the descriptive features is the percentage of the within-cluster similarity that this particular feature can explain. For example, for the 0th cluster, the 5mer "GTSMA" explains 58.5% of the average similarity between the objects of the 0th cluster. A similar quantity is displayed for each of the discriminating features and is the percentage of the dissimilarity between the cluster and the rest of the objects that this feature can explain. In general, there is a large overlap between descriptive and discriminating features, with the only difference being that the percentages associated with the discriminating features are typically smaller than the corresponding percentages of the descriptive features. This is because some of the descriptive features of a cluster may also be present in a small fraction of the objects that do not belong to the cluster.

6.6. Visualizing the Clusters

CLUTO's programs can produce a number of visualizations that can be used to see the relationships among the clusters, objects, and features. You have already seen one of them in **Fig. 2** that was produced by the -plotmatrix parameter. The same parameter can be used to visualize sparse high-dimensional data sets. This is illustrated in **Fig. 4A** for the protein data set used earlier. As we can see from that plot, vcluster shows the rows of the input matrix reordered in such a way that the rows assigned to each of the 10 clusters are numbered consecutively. The columns of the displayed matrix are selected to be the union of the most descriptive and discriminating features of each cluster and are ordered according to the tree produced by an agglomerative clustering of the columns. In addition, at the top of each column, the label of each feature is shown. Each nonzero positive element of the matrix is displayed by a different

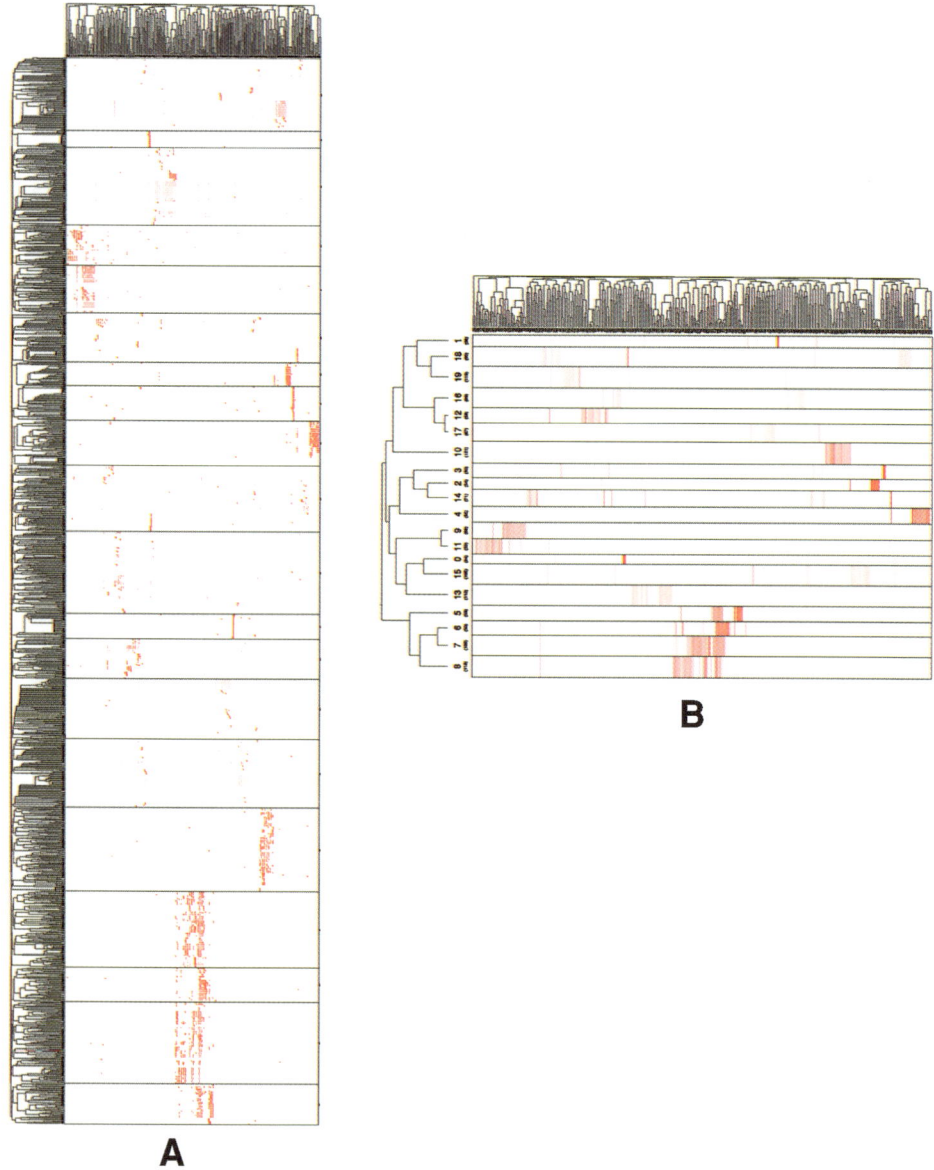

Fig. 4. Various visualizations generated by the -plotmatrix (**A**) and -plotclusters (**B**) parameters.

shade of red. Entries that are bright red correspond to large values and the brightness of the entries decreases as their value decreases. Also note that in this visualization both the rows and columns have been reordered using a hierarchical tree.

Finally, **Fig. 4B** shows the type of visualization that can be produced when -plotcluster is specified for a sparse matrix. This plot shows the clustering solution shown in **Fig. 4A** by replacing the set of rows in each cluster by a single row that corresponds to the centroid vector of the cluster. The -plotcluster option is particularly useful for displaying very large data sets, since the number of rows in the plot is only equal to the number of clusters.

References

1. Linial, M., Linial, N., Tishby, J., and Golan, Y. (1997) Global self-organization of all known protein sequences reveals inherent biological structures. *J. Mol. Biol.* **268,** 539–556.
2. Enright, A. J. and Ouzounis, C. A. (2000) GeneRAGE: a robust algorithm for sequence clustering and domain detection. *Bioinformatics* **16(5),** 451–457.
3. Mian, I. S. and Dubchak, I. (2000) Representing and reasoning about protein families using generative and discriminative methods. *J. Comput. Biol.* **7(6),** 849–862.
4. Spang, R. and Vingron, M. (2001) Limits of homology of detection by pairwise sequence comparison. *Bioinformatics* **17(4),** 338–342.
5. Kriventseva, E., Biswas, M., and Apweiler, R. (2001) Clustering and analysis of protein families. *Curr. Opin. Struct. Biol.* **11,** 334–339.
6. Eidhammer, I., Jonassen, I., and Taylor, W. R. (2000) Structure comparison and stucture patterns. *J. Comput. Biol.* **7,** 685–716.
7. Shatsky, M., Fligelman, Z. Y., Nussinov, R., and Wolfson, H. J. (2000) Alignment of flexible protein structures, in *Proceedings of the 8th International Conference on Intelligent Systems for Molecular Biology*, (Altman, R., et al., eds.), AAAI Press, Menlo Park, CA, pp. 329–343.
8. Lee, R. C. T. (1981) Clustering analysis and its applications, in *Advances in Information Systems Science* (Toum, J. T., ed.), Plenum, New York.
9. Jain, A. K. and Dubes, R. C. (1988) *Algorithms for Clustering Data*, Prentice Hall, 1998.
10. Cheeseman, P. and Stutz, J. (1996) Bayesian classification (autoclass): theory and results, in *Advances in Knowledge Discovery and Data Mining* (Fayyad, U. M., Piatetsky-Shapiro, G., Smith, P., and Uthurusamy, R., eds.), AAAI/MIT Press, pp. 153–180.
11. Kaufman, L. and Rousseeuw, P. J. (1990) *Finding Groups in Data: An Introduction to Cluster Analysis*, John Wiley & Sons, New York.
12. Jackson, J. E. (1991) *A User's Guide to Principal Components*, John Wiley & Sons, New York.

13. Ng, R. and Han, J. (1994) Efficient and effective clustering method for spatial data mining, in *Proceedings of the 20th VLDB Conference* (Bocca, J. B., Jarke, M., and Zaniolo, C., eds.), San Francisco, CA, pp. 144–155.
14. Berry, M. W., Dumais, S. T., and O'Brien, G. W. (1995) Using linear algebra for intelligent information retrieval. *SIAM Rev.* **37,** 573–595.
15. Zhang, T., Ramakrishnan, R., and Linvy, M. (1996) Birch: an efficient data clustering method for large databases, in *Proceedings of 1996 ACM-SIGMOD International Conference on Management of Data*, ACM Press, New York, NY, pp. 103–114.
16. Ester, M., Kriegel, H.-P., Sander, J., and Xu, X. (1996) A density-based algorithm for discovering clusters in large spatial databases with noise, in *Proceedings of the Second International Conference on Knowledge Discovery and Data Mining* (Simoudis, E., Han, J., and Fayyad, U., eds.), AAAI Press, Menlo Park, CA, pp. 226–231.
17. Wang, X., Wang, J. T. L., Shasha, D., Shapiro, B., Dikshitulu, S., Rigoutsos, I., and Zhang, K. (1997) Automated discovery of active motifs in three dimensional molecules, in *Proceedings of the 3rd International Conference on Knowledge Discovery and Data Mining* (Heckerman, D., Mannila, H., Pregibon, D., and Uthurusamy, R., eds.), AAAI Press, Menlo Park, CA, pp. 89–95.
18. Guha, S., Rastogi, R., and Shim, K. (1998) CURE: an efficient clustering algorithm for large databases, in *Proceedings of 1998 ACM-SIGMOD International Conference on Management of Data*, ACM Press, New York, NY, pp. 73–84.
19. Guha, S., Rastogi, R., and Shim, K. (1999) ROCK: a robust clustering algorithm for categorical attributes, in *Proceedings of the 15th International Conference on Data Engineering*, IEEE, Washington-Brussels-Tokyo, pp. 512–521.
20. Karypis, G., Han, E. H., and Kumar, V. (1999) Chameleon: a hierarchical clustering algorithm using dynamic modeling. *IEEE Comput.* **32(8),** 68–75.
21. Jain, A. K., Murty, M. N., and Flynn, P. J. (1999) Data clustering: a review. *ACM Comput. Surv.* **31(3),** 264–323.
22. Han, J., Kamber, M., and Tung, A. K. H. (2001) Spatial clustering methods in data mining: a survey, in *Geographic Data Mining and Knowledge Discovery* (Miller, H. and Han, J., eds.), Taylor & Francis.
23. MacQueen, J. (1967) Some methods for classification and analysis of multivariate observations, in *Proceedings of the 5th Symposium Math. Statist. Prob.*, pp. 281–297.
24. Dempster, A. P., Laird, N. M., and Rubin, D. B. (1977) Maximum likelihood from incomplete data via the em algorithm. *J. Roy. Stat. Soc.* **39**.
25. Zahn, K. (1971) Graph-theoretical methods for detecting and describing gestalt clusters. *IEEE Trans. Comput.* **(C-20),** 68–86.
26. Han, E. G., Karypis, G., Kumar, V., and Mobasher, B. (1998) Hypergraph based clustering in high-dimensional data sets: a summary of results. *Bull. Tech. Committee Data Eng.* **21(1)**.
27. Strehl, A. and Ghosh, J. (2000) Scalable approach to balanced, high-dimensional clustering of market-baskets, in *Proceedings of HiPC*, Springer Verlag, pp. 525–536.

28. Boley, D. (1998) Principal direction divisive partitioning. *Data Mining Knowl. Discov.* **2(4)**,
29. Duda, R. O., Hart, P. E., and Stork, D. G. (2001) *Pattern Classification*, John Wiley & Sons, New York.
30. Zhao, Y. and Karypis, G. (2001) Criterion functions for document clustering: experiments and analysis. Technical report TR #01–40, Department of Computer Science, University of Minnesota, Minneapolis. Available at http://cs.umn.edu/~karypis/publications.
31. Sneath, P. H. and Sokal, R. R. (1973) *Numerical Taxonomy*, Freeman, London, UK.
32. King, B. (1967) Step-wise clustering procedures. *J. Am. Stat. Assoc.* **69**, 86–101.
33. Johnson, M. S., Sutcliffe, M. J., and Blundell, T. L. (1990) Molecular anatomy: phyletic relationships derived from 3-dimensional structures of proteins. *J. Mol. Evol.* **30**, 43–59.
34. Taylor, W. R., Flores, T. P., and Orengo, C. A. (1994) Multiple protein structure alignment. *Protein Sci.* **3**, 1858–1870.
35. Jonyer, I., Holder, L. B., and Cook, D. J. (2000) Graph-based hierarchical conceptual clustering in structural databases. *Int. J. Artific. Intell. Tools.*
36. Eddy, S. R. (1996) Hidden markov models. *Curr. Opin. Struct. Biol.* **6**, 361–365.
37. Eddy, S. R. (1998) Profile hidden markov models. *Bioinformatics* **14**, 755–763.
38. Aggarwal, C. C., Gates, S. C., and Yu, P. S. (1999) On the merits of building categorization systems by supervised clustering, *Proceedings of the Fifth ACM SIGKDD International Conference on Knowledge Discovery and Data Mining* (Chaudhuri, S. and Madigan, D., eds.), ACM Press, New York, NY, pp. 352–356.
39. Cutting, D. R., Pedersen, J. O., Karger, D. R., and Tukey, J. W. (1992) Scatter/gather: a cluster-based approach to browsing large document collections, in *Proceedings of the ACM SIGIR*, ACM Press, New York, NY, pp. 318–329.
40. Larsen, B. and Aone, C. (1999) Fast and effective text mining using linear-time document clustering, in *Proceedings of the Fifth ACM SIGKDD International Conference on Knowledge Discovery and Data Mining* (Chaudhuri, S. and Madigan, D., eds.), ACM Press, New York, NY, pp. 16–22.
41. Steinbach, M., Karypis, G., and Kumar, V. (2000) A comparison of document clustering techniques, in *KDD Workshop on Text Mining*, ACM Press, New York, NY, pp. 109–110.
42. Agrawal, R., Gehrke, J., Gunopulos, D., and Raghavan, P. (1998) Automatic subspace clustering of high dimensional data for data mining applications, in *Proceedings of 1998 ACM-SIGMOD International Conference on Management of Data*, ACM Press, New York, NY, pp. 94–105.
43. Burdick, D., Calimlim, M., and Gehrke, J. (2001) Mafia: a maximal frequent itemset algorithm for transactional databases, in *Proceedings of the 17th International Conference on Data Engineering*, IEEE, Washington-Brussels-Tokyo, pp. 443–452.
44. Nagesh, H., Goil, S., and Choudhary, A. (1999) Mafia: efficient and scalable subspace clustering for very large data sets. Technical report TR #9906-010, Northwestern University.

45. Huang, Z. (1997) A fast clustering algorithm to cluster very large categorical data sets in data mining, in *Proceedings of SIGMOD Workshop on Research Issues on Data Mining and Knowledge Discovery*, Tucson, Arizona.
46. Ganti, V., Gehrke, J., and Ramakrishnan, R., (1999) Cactus-clustering categorical data using summaries, in *Proceedings of the 5th International Conference on Knowledge Discovery and Data Mining* (Chaudhuri, S. and Madigan, D., eds.), ACM Press, New York, NY, pp. 73–83.
47. Gibson, D., Kleinberg, J., and Rahavan, P. (1998) Clustering categorical data: an approach based on dynamical systems, in *Proceedings of the 24th International Conference on Very Large Databases* (Gupta, A., Shmueli, O., and Widom, J., eds.), pp. 311–323, San Francisco, CA, Morgan Kaufmann.
48. Ryu, T. W. and Eick, C. F. (1998) A unified similarity measure for attributes with set or bag of values for database clustering, in *The 6th International Workshop on Rough Sets, Data Mining and Granular Computing*, Research Triangle Park, NC.
49. Gusfield, D. (1997) *Algorithms on Strings, Trees, and Sequences: Computer Science and Computational Biology*, Cambridge University Press, NY.
50. Dayhoff, M. O., Schwartz, R. M., and Orcutt, B. C. (1978) A model of evolutionary change in proteins. *Atlas Protein Sequence Struct.* **5,** 345–352.
51. Schwartz, R. M. and Dayhoff, M. O. (1978) Matrices for detecting distant relationships. *Atlas Protein Sequence Struct.* **5,** 353–358.
52. Henikoff, S. and Henikoff, J. G. (1992) Amino acid substitution matrices from protein blocks. *Proc. Natl. Acad. Sci. USA* **89,** 10,915–10,919.
53. Needleman, S. B. and Wunsch, C. D. (1970) A general method applicable to the search for similarities in the amino acid sequence of two proteins. *J. Mol. Biol.* **48,** 443–453.
54. Smith, T. F. and Waterman, M. S. (1981) Identification of common molecular subsequences. *J. Mol. Biol.* **147,** 195–197.
55. Lipman, D. J. and Pearson, W. R. (1985) Rapid and sensitive protein similarity searches. *Science* **227,** 1435–1441.
56. Pearson, W. R. and Lipman, D. J. (1988) Improved tools for biological sequence comparison. *Proc. Natl. Acad. Sci. USA* **85,** 2444–2448.
57. Altschul, S., Gish, W., Miller, W., Myers, E., and Lipman, D. (1990) Basic local alignment search tool. *J. Mol. Biol.* **215,** 403–410.
58. Yona, G., Linial, N., and Linial, M. (2000) Protomap: automatic classification of protein sequences and hierarchy of protein families. *Nucleic Acids Res.* **28,** 49–55.
59. Bolten, E., Schliep, A., Schneckener, S., Schomburg, D., and Schrader, R. (2001) Clustering protein sequences-structure prediction by transitive homology. *Bioinformatics* **17,** 935–941.
60. Johnson, M. S. and Lehtonen, J. V. (2000) Comparison of protein three dimensional structure, in *Bioinformatics: Sequences, Structure and Databanks* (Higgins, D, eds.), Oxford University Press, 2000.
61. Grindley, H. M., Artymiuk, P. J., Rice, D. W., and Willet, P. (1993) Identification of tertiary structure resemblance in proteins using a maximal common subgraph isomorphism algorithm. *J. Mol. Biol.* **229,** 707–721.

62. Mitchell, E. M., Artymiuk, P. J., Rice, D. W., and Willet, P. (1989) Use of techniques derived from graph theory to compare secondary structure motifs in proteins. *J. Mol. Biol.* **212,** 151–166.
63. Holm, L. and Sander, C. (1993) Protein structure comparison by alignment of distance matrices. *J. Mol. Biol.* **233,** 123.
64. Kleywegt, G. J. and Jones, T. A. (1997) Detecting folding motifs and similarities in protein structures. *Methods Enzymol.* **277,** 525–545.
65. Calinski, T. and Harabasz, J. (1974) A dendrite method for cluster analysis. *Commun. Stat.* **3,** 1–27.
66. Ben-Hur, A., Elisseeff, A., and Guyon, I. (2000) A stability based method for discovering structure in clustered data, in *Pacific Symposium on Biocomputing,* vol. 7, World Scientific Press, Singapore, pp. 6–17.
67. Yeung, K. Y., Haynor, D. R., and Ruzzo, W. L. (2001) Validating clustering for gene expression data. *Bioinformatics* **17(4),** 309–318.
68. Fodor, S. P., Rava, R. P., Huang, X. C., Pease, A. C., Holmes, C. P., and Adams, C. L. (1993) Multiplexed biochemical assays with biological chips. *Nature* **364,** 555, 556.
69. Schena, M., Shalon, D., Davis, R. W., and Brown, P. O. (1995) Quantitative monitoring of gene expression patterns with a complementary DNA microarray. *Science* **270.**
70. Brazma, A., Robinson, A., Cameron, G., and Ashburner, M. (2000) One-stop shop for microarray data. *Nature* **403,** 699–701.
71. Yang, Y. H., Dudoit, S., Luu, P., and Speed, T. P. (2001) Normalization for cDNA microarray data, in *Proceedings of SPIE International Biomedical Optics Symposium* (Bittner, M. L., Chen, Y., Dorsel, A. N., and Dougherty, E. R., eds.), vol. 4266, SPIE, Bellingham, WA, pp. 141–152.
72. D'Haeseleer, P., Fuhrman, S., Wen, X., and Somogyi, R. (1998) Mining the gene expression matrix: inferring gene relationships from large scale gene expression data, in, *Information Processing in Cells and Tissues,* Plenum, New York, pp. 203–212.
73. Eisen, M. (2000) Cluster 2.11 and treeview 1.50 http://rana.lbl.gov/Eisen Software.htm.
74. Tamayo, P., Slonim, D., Mesirov, J., Zhu, Q., Kitareewan, S., Dmitrovsky, E., Lander, E. S., and Golub, T. R. (1999) Interpreting patterns of gene expression with self-organizing maps: methods and application to hematopoietic differentiation. *Proc. Natl. Acad. Sci. USA* **96,** 2907–2912.
75. Kohonen, T. *Self-Organizing Maps,* Springer-Verlag, Berlin, 1995.
76. Ben-Dor, A. and Yakhini, Z. (1999) Clustering gene expression patterns, in *Proceedings of the 3rd Annual International Conference on Research in Computational Molecular Biology,* ACM Press, New York, NY, pp. 33–42.
77. Shamir, R. and Sharan, R. (2000) Click: a clustering algorithm for gene expression analysis, in *Proceedings of the 8th International Conference on Intelligent Systems for Molecular Biology* (Altman, R., et al., eds.), AAAI Press, Menlo Park, CA, pp. 307–316.

78. Zhao, Y. and Karypis, G. (2002) Comparison of agglomerative and partitional document clustering algorithms, in *SIAM (2002) Workshop on Clustering High-Dimensional Data and Its Applications*. Arlington, VA. Also available as technical report #02-014, University of Minnesota.
79. Karypis, G. and Kumar, V. (1998) hMETiS 1.5: a hypergraph partitioning package. Technical report, Department of Computer Science, University of Minnesota. Available at www-users.cs.umn.edu/~karypis/metis
80. Karypis, G. and Kumar, V. (1998) METiS 4.0: unstructured graph partitioning and sparse matrix ordering system. Technical report, department of Computer Science, University of Minnesota. Available at www-users.cs.umn.edu/~karypis/metis

14

A Primer on the Visualization of Microarray Data

Paul Fawcett

1. Introduction

DNA microarrays represent a powerful technology offering unprecedented scope for discovery *(1)*. However, the ability to measure, in parallel, the gene expression patterns for thousands of genes represents both the strength and a key weakness of microarrays. One of the central challenges of functional genomics has been to cope with the enormity of microarray data sets, and, indeed, the usefulness of microarrays has been limited by our ability to extract useful information from these data. In general terms, analyzing microarray data requires a series of numerical transformations and/or filters intended to extract from the data set the subset of represented genes that may be of interest. The resulting lists generally represent genes with large variance or periodicity within their gene expression vectors *(2)*; high fold inductions over a time course *(3)*; genes that are considered significant by some statistical criterion *(4)*; or genes that meet some other threshold, such as exceeding a given percentile rank in the distribution of ratios *(5,6)*. However, examining a spreadsheet of gene names and expression ratios often provides little insight into the interesting trends or patterns that may exist within the data. Rather, methods have been developed for both the classification and display of these data sets. Indeed, given the non-hypothesis-driven nature of many microarray experiments, the ability to readily visualize trends in the data assumes paramount importance.

The goal of this chapter is therefore to provide an overview and introduction to some of the software packages developed in the Brown and Botstein laboratories at Stanford University for the visual display of microarray data, and to introduce the reader to these methods using straightforward and nonmathematical language. It is not our goal to discuss the previously mentioned transformations or filters required to determine the genes of interest

From: *Methods in Molecular Biology: vol. 224: Functional Genomics: Methods and Protocols*
Edited by: M. J. Brownstein and A. Khodursky © Humana Press Inc., Totowa, NJ

but, rather, to discuss how data may be explored once these preliminary steps have been applied. Also introduced, for the first time, is a new tool, DecCor2, designed to allow the genome-order display of aggregate microarray data.

1.1. A Note on Gene Expression Data

Array data are invariably expressed in the form of ratios, so a preliminary discussion is required to indicate how these are normally represented. When expression ratios are shown numerically, they are most commonly expressed in logarithmic space, typically to the base 2, i.e., ratio = $\log_2$(red intensity/green intensity). The resulting ratios have the property of being symmetrical about zero, and a negative sign denotes ratios with a larger denominator (making it much easier to grasp the size of such ratios, since these ratios are no longer compressed between 0 and 1, as occurs in real space). Although the base used is not critical, base 2 logarithms are commonly employed because ratios thus expressed are easily converted to "fold inductions" in real space, owing to the familiarity of powers of two as used in the binary numbering system (i.e., a $\log_2$-transformed ratio of $4 = 2^4$ = a 16-fold in difference in real space). Note that logarithms to the base 10 are seldom employed, for the simple reason that most biologically relevant ratios tend to fall in the 1- to 10- or 1- to 100-fold range. However, even when ratios are log transformed, it is still difficult to make sense of columns of numbers, and therefore sensible alternatives are required.

2. TreeView and Cluster

One of the the most widely used programs for displaying gene expression data is TreeView, developed by Michael Eisen, and available on the World Wide Web at http://rana.lbl.gov, or at http://microarrays.org. TreeView was the first program to exploit the idea that gene expression ratios can be easily and intuitively expressed colorimetrically *(7)*, a notion that has gained general currency. As shown in **Fig. 1**, the genes in the input set are ordered as the rows

---

Fig. 1. *(see facing page)* Output from the Cluster/Treeview suite, courtesy of Arkady Khodursky. Rows correspond to individual genes, which are named in the column on the right, while each colorized column corresponds to separate *Escherichia coli*–specific arrays. In this example, data from 120 *E. coli* genes have been extracted from 40 arrays. These arrays correspond to five discrete time courses, each representing a different treatment. As explained in the text, gene expression ratios have been rendered colorimetrically, with green shading indicating decreased expression relative to time zero, and red shading indicating increased expression relative to time zero. Gray shading indicates missing or unreliable data. The dendrogram on the left recapitulates the pairing of genes and nodes assigned during hierarchical clustering. The length of the branches is proportional to the distance between the nodes, as is best illustrated by the very deep branch splitting the two penultimate nodes that intuitively divide this data set.

Primer on the Visualization of Microarray Data

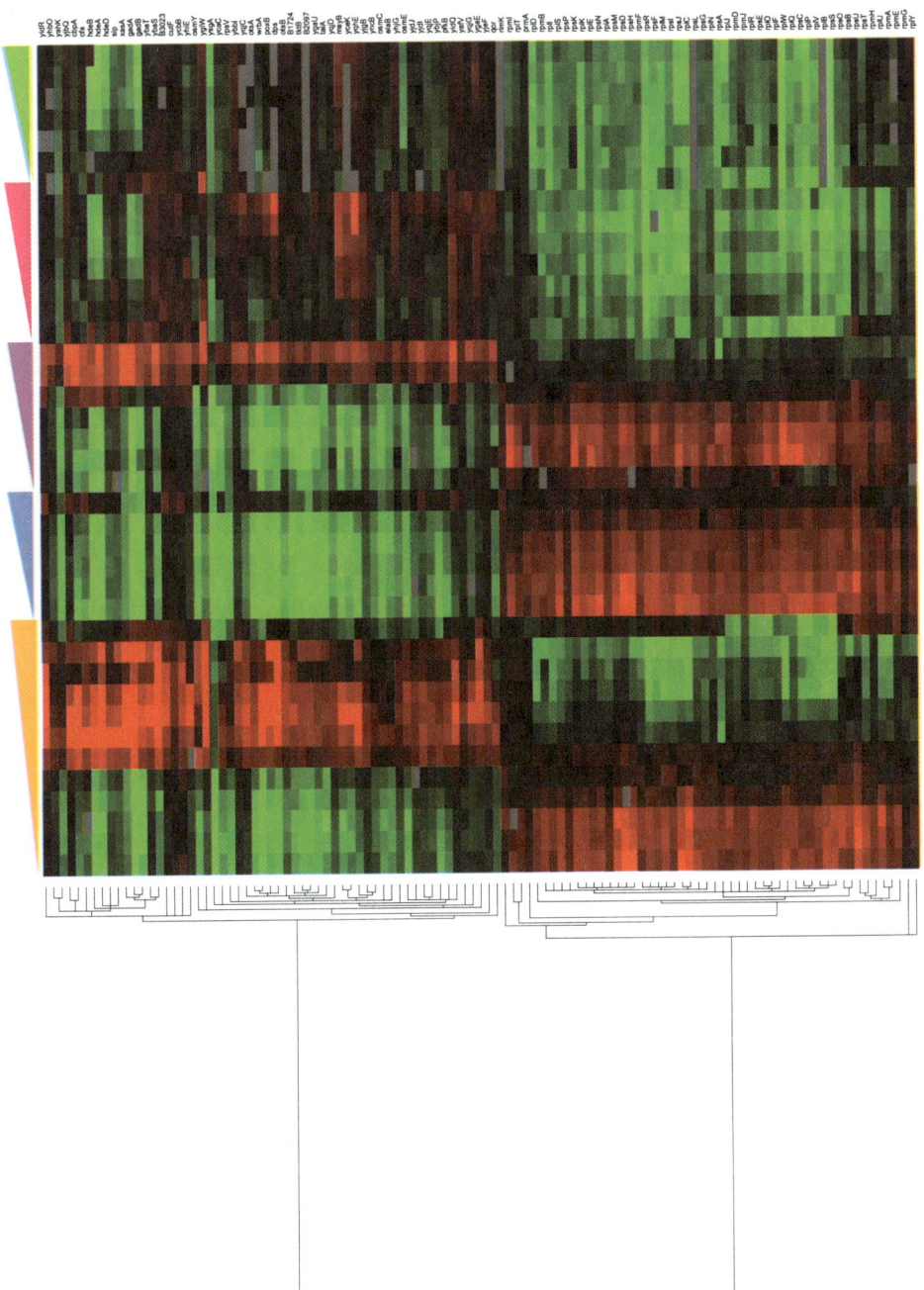

of the figure, while arrays are arranged in columns. When reading across a row from left to right, one is therefore looking at a colorized depiction of the gene expression vector for a particular gene, representing the behavior of a given gene's expression over the set of conditions represented by the arrays (columns). In this colorization scheme, more intense hues are used to denote more extreme ratios. This intuitively recapitulates what is seen on a pseudo-colored photomicrograph of a scanned glass-substrate array hybridized with two different fluorescent probes (typically Cy3 and Cy5 fluorophores, which produce, respectively, green and red signals). However, while a bright red band in TreeView indeed corresponds to an excess of Cy5 relative to Cy3, and a bright green band denotes the reverse, TreeView uses shades close to black as the log ratio of Cy3 to Cy5 approaches 0, although this would appear yellow on an array. One must also bear in mind that since TreeView uses ratio data, any information regarding the absolute signal intensity in either channel is lost (i.e., Cy5 = 65,535/Cy3 = 65,535 and Cy5 = 32/Cy3 = 32 appear to be equivalent). Care must therfore be taken when filtering the data to ensure that the calculated ratios are meaningful and not mearly a reflection of variability that exists below the threshold of reliable detection (i.e., a reflection of the noise present in very low-level signals). Note that while TreeView is designed for displaying gene expression data, it is not intended for classification and is most usually used in conjunction with its companion program, Cluster, which (among other things) can produce an input file suitable for use with TreeView using hierarchical clustering *(7)*, (Chapter 13).

The output of the TreeView program is a display of colorized gene expression ratios along with a dendrogram that reflects the degree of similarity between individual genes as well as the order in which genes were assembled into clusters by means of the algorithm in Cluster. This plot is commonly referred to as a "clustergram" or an "Eisen plot." It has often been demonstrated that genes with similar expression profiles that cluster together tend to be functionally or biochemically related, and often represent genes that work together in a pathway *(3)*. While this is not a hard-and-fast rule, co-clustering (especially in the context of a large input data set) is suggestive that closely clustered genes may have regulatory commonalities. It should additionally be pointed out that, although the order of the columns representing arrays is often specified by the user (as when a time course or dosage series is presented), it is also possible to use hierarchical clustering to order the arrays. This is often used to classify source tissues (e.g., as normal or cancerous). For instance, clustering of this type has been employed to help demonstrate that diffuse large B-cell lymphoma has molecularly distinct variants with different outcomes, and that these variants can be distinguished using this approach *(8)*.

The technical details for using Cluster and TreeView are well covered elsewhere and are not discussed here. However, note that Cluster allows the user to select a preferred similarity metric from a range of options, as well as to select variations of the clustering process. In practice, the best course is to experiment with the available options in order to generate a clustergram that appears to present the data in the most coherent fashion; there is no universally "best" method for clustering. As a final caveat, it should be understood that the final dendrogram produced by TreeView represents only one of the many valid dendrograms that could be produced by "flipping" the branches of the dendrogram at any node.

3. Promoter

Developed by Joe DeRisi, now at UCSF, Promoter is a package designed to display features associated with genomes and to allow the user to perform powerful searches to retrieve associated sequence data. This package is available at both http://microarrays.org and the DeRisi laboratory Web site at http://derisilab.ucsf.edu. The display feature of Promoter is particularly relevant because it creates genome maps with features annotated in genome order, and with a correct scale representation of both the size and position of the feature on the genome or chromosome of interest. Features are also correctly displayed on either the Crick or Watson strands or can be marked as intergenic. By loading an optional "gene data" file (consisting simply of a tab-delimited text file with feature names and an associated log ratio, with each feature separated with a carriage return), Promoter also has the ability to map onto the features data a colorimetric representation of array data. Ultimately, Promoter produces output in PostScript format suitable for direct importation in Adobe Illustrator or similar vector-drawing software, where the annotation is then often customized to suit the figure being created. An excerpt from a recently published example is shown in **Fig. 2**, which illustrates the results for a single yeast chromosome for a chromosomal immunoprecipitation array experiment performed using antibodies against the Rap1 transcription factor (see Chapter 8 for a description of chromosomal immunoprecipitation on chip methodology).

Although not a focus of this discussion, it should also be noted that Promoter's ability to extract primary sequence data from genetic regions corresponding to hits in microarray experiments is an important preliminary step when attempting to correlate array results with features of the primary sequence data, such as when performing scans for the presence of possible regulatory motifs. Promoter also has the ability to extract primary sequence-containing motifs specified by the user using the IUPAC degenerate nucleotide code, with a user-selected acceptable number of mismatches from consensus.

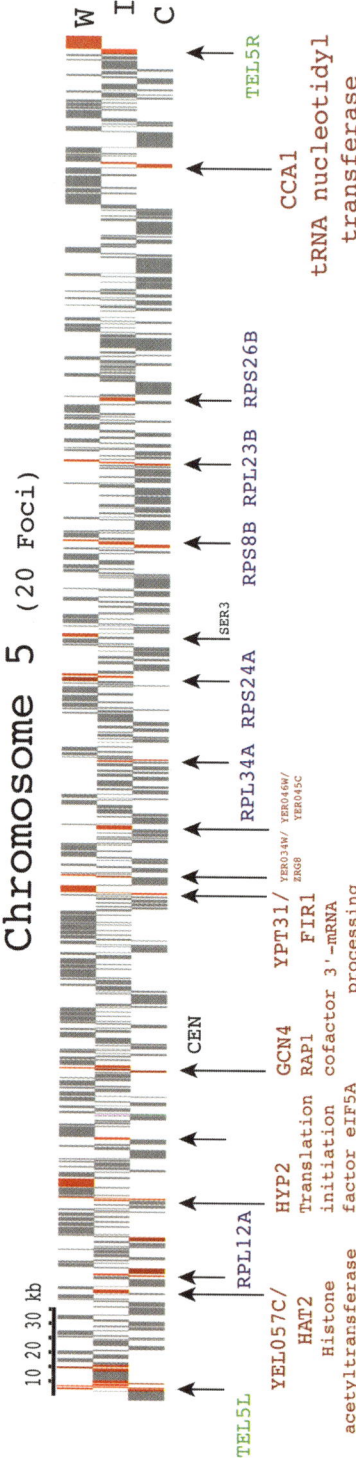

Fig. 2. Output from Promoter, with additional annotation created in Adobe Illustrator, courtesy of Jason Lieb. This figure illustrates chromosome 5 of *Saccharomyces cerevisiae*, with each feature shown in genome order, and sized proportionately. The top and bottom rows indicate, respectively, genes on the Watson and Crick strands, while the middle bar indicates regions marked as intergenic. This figure was generated using data from a chromosomal immunoprecipitation array experiment designed to determine the targets of the Rap1 transcription factor, and these target genes have been colored red. Note that when used in conjunction with standard gene expression data, Promoter permits full red/green colorization of expression ratios.

4. Caryoscope

Unlike the other display methods discussed, Caryoscope (available at http://genome-www.stanford.edu/~rees/), developed by Christian Rees of the Botstein laboratory at Stanford, is not primarily intended to summarize and display gene expression data but, rather, to display the results of array comparative genomic hybridizations (aCGH). The essential idea of array-based CGH is that, instead of cDNA produced from mRNA, genomic DNA (gDNA) can be directly used in array experiments. Comparative gDNA hybridization can therefore be used for gene copy estimation *(9)*. As with gene expression studies, two genomic DNA samples are hybridized at once to the same array, each having been labeled with a different fluorophore. Typically, these genomic DNA samples are obtained from either a normal and a diseased tissue (i.e., channel one measures a sample from a cancerous tissue, which may have undergone gene duplications or deletions, while channel two measures a sample from a normal tissue) *(9,10)*, or from two closely related species, such as recently evolutionarily diverged strains of yeast or bacteria *(11,12)*. In either case, obtaining a log ratio that significantly deviates from zero is presumptive evidence that either a gene amplification has occurred in one of the tissues, or that the gene copy number has been increased in the other tissue. Caryoscope is designed to display these data in a genome-ordered fashion. Separate versions are available for both yeast and human physical map data. The ability to display aCGH ratio data in the context of a physical map is particularly useful given that deletions or duplications often occur on a larger scale than individual genes; indeed, amplifications or deletions often occur in large contiguous regions, and this is immediately evident on examination of the data with Caryoscope.

A very nice feature of Caryoscope is that a separate diagram is produced for each of the chromosomes of the organism of interest. As shown in **Fig. 3**, Caryoscope displays ratios of greater than zero as red bars to the right of the line representing the chromosome, while green bars representing ratios of less than zero are displayed to the left of the line. In this case, color is used only to indicate the sign of the ratio, while the actual deviation of the log ratio from zero is proportional to the height of the bar representing a particular locus. In the example shown, the lines drawn to the right of the main line indicate the ratio associated with a hypothetical bar of that length. Clearly, most of the bars in **Fig. 3** are of a length consistent with fluctuation within the "noise" of the expression ratios, but with notable peaks corresponding to areas of copy number variation.

5. DecCor2

DecCor2 is a new software package for displaying summaries of gene expression in genome order. Unlike Caryoscope, or an earlier genome-order

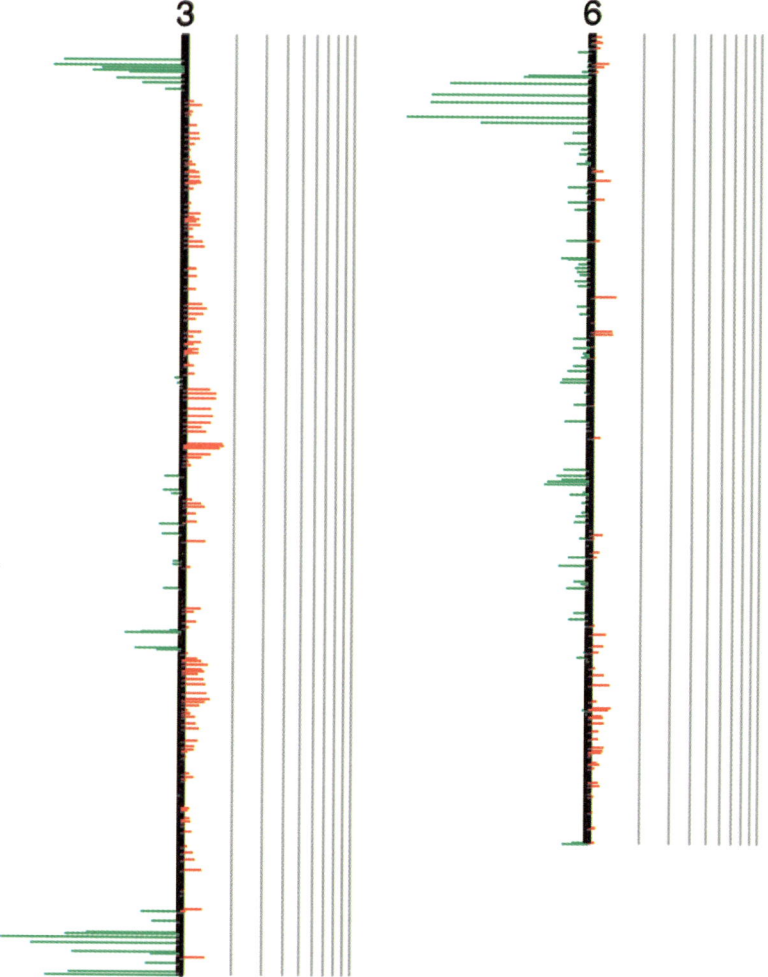

Fig. 3. Output from Caryoscope, courtesy of Barbara Dunn. This figure illustrates the results for yeast chromosomes 3 and 6 of a CGH carried out between a strain of yeast used for brewing ale (red) and a standard laboratory strain (green). In this example, the value indicated by the length of the bars is based on a value determined by a sliding window of size 3 representing the average ratio of nearby features. The length of the bars is proportional to the red/green channel ratio, and the logarithmic scale bars drawn to the right of the line denoting the chromosome can be used to gage the magnitude of this ratio. In this example, most deviations appear to be within the range of experimental noise, with the exception of the obvious strong deviations near the telomeric regions.

display program developed by Dan Zimmer of Sydney Kustu's laboratory (University of California at Berkeley) that rearranges the actual image spots from array photomicrographs *(13)*, DecCor2 colorimetrically displays correlation coefficient data calculated from many array experiments. The unique feature of this program is therefore that the colorimetric encoding confers information about the similarity of the gene expression pattern of a gene of interest to that of all other genes represented on an array. DecCor2 additionally provides nongraphic output for each gene in the input data set indicating the group of genes that shares the highest degree of correlation or anticorrelation.

DecCor2 is designed to distill very large array data sets into a form that quickly allows the identification of those genes with an expression profile most similar (or most dissimilar) to the expression profile of a gene of interest. Indeed, the reliability of the correlation coefficients will tend to increase with the number of available arrays (presumably corresponding to an increasing number of conditions). Since this information can optionally be presented in a structured fashion (such as genome ordering), spatial or functional relationships among genes of interest can be made immediately apparent. Spatial relationships often hint at possible operon structure, or relationship of the gene of interest's expression pattern to the general cellular growth rate as seen by correlation to ribosomal proteins or other hallmarks of the general cellular growth rate. Often DecCor2 analysis will reveal that the expression pattern of a gene of interest correlates with a broader gene expression program of known genes involved in a common process, such as motility or chemotaxis, or shift to stationary phase.

DecCor2 works by calculating a matrix of similarity metrics representing all of the possible pairwise combinations of genes in the input data set. The similarity metric may be specified by the user, and either the Pearson correlation coefficient or the uncentered variant of the Pearson presented by Michael Eisen may be chosen. Unlike TreeView, the colors do not represent induction ratios but, rather, the similarity metric calculated using the entire available gene expression vector (**Fig. 4**). Bright red shades now indicate a correlation coefficient approaching or equal to 1, while bright green shades indicate a strong anticorrelation approaching or equal to −1. Unlike the dendrograms produced by hierarchical clustering, the output of DecCor2 is intended to be displayed in genome order. Every gene in the input data set will have its own genome-order representation of similarity metrics (i.e., gene of interest vs all other genes). Therefore, provision is made within DecCor2 both to browse through the maps corresponding to each gene in the input set and to search directly for genes of interest.

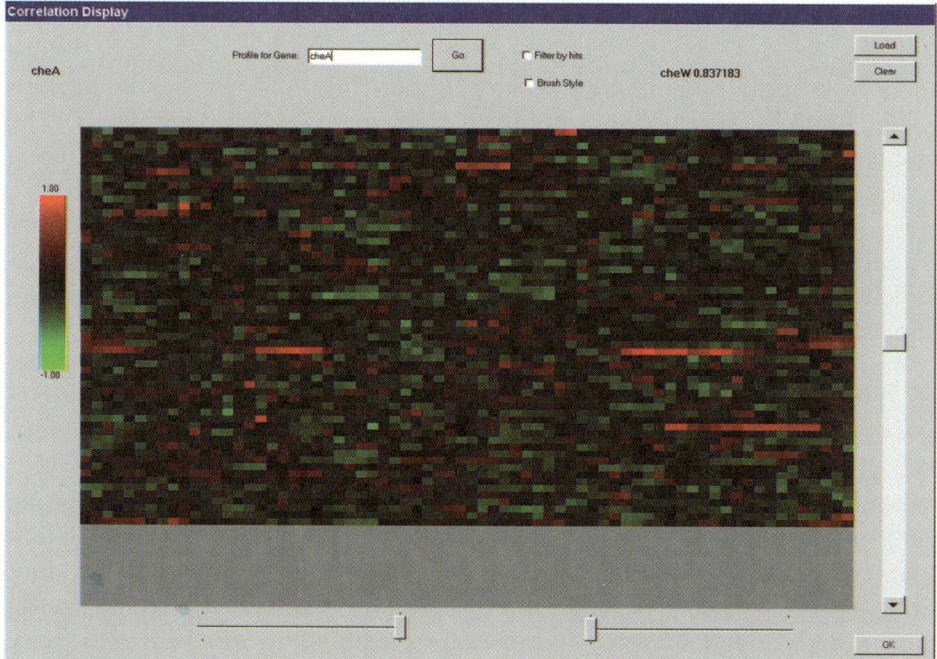

Fig. 4. Sceen shot from DecCor2. This typical screen shot shows the correlation coefficients for the *E. coli cheA* gene, relative to all other *E. coli* genes, as represented in genome order, and displayed colorimetrically as described in the text. Note that in several regions, long contiguous regions of genes appear to be coexpressed with *cheA*, as indicated by bright red coloration. In this example, data from more than 200 *E. coli*–specific arrays (courtesy of A. Khodursky), representing dozens of experimental conditions, were used for calculating the correlation coefficients. Refer to **Subheading 5.3.** for a description of the software controls.

Aside from the visual display, DecCor2 exploits the characteristics of the distribution of gene expression vector similarities in order to output lists of genes representing outliers in the distribution. This allows the easy creation of lists of genes most strongly similar in their expression pattern to a gene of interest. By specifying the number of standard deviations (SDs) away from the mean of the distribution that a gene should be before being included in an output list, the user has direct control over the inclusion parameter. Either the distribution of gene-specific similarities (i.e., a gene-specific cutoff) or the global distribution generated by considering all pairwise similarities (i.e., a global cutoff) may be used.

5.1. Equipment Requirements

DecCor2 was developed and tested on a PC-compatible machine running Microsoft Windows NT 4.0, but also runs correctly under Windows 2000 and Windows XP. There is a known problem with Windows 95/98 when searching for a gene profile by name: it will not allow entry of the gene name. This software requires a large amount of memory and 512 megabytes or more of random access memory is recommended.

5.2. Data Formats for DecCor2

DecCor2 minimally requires a data input file and may optionally use a second file format that allows the user to highlight specific user-specified sets of genes.

5.2.1. Array Data

The first input file that is required is the transcriptional profiling data. This should be a tab-delimited text format file (which may conveniently be created using the "Save As" feature from within Microsoft Excel or another spreadsheet package). Each row of this file represents the gene expression vector for a discrete spot on the arrays. The first column should therefore be a text identifier for the spot (this will in most cases be a gene name). The user must ensure that each row has a unique identifier, or the behavior of the program will be undefined. Additionally, there must be only one such column of identifiers. Future versions should support the inclusion of an optional description column, but this is not supported at this time. Note that the order in which the gene names appear in the first column determines the order in which the genes will be displayed on the screen. If you wish to use the program to display data in genome order, this order must be reflected in the order of the genes in the input file. Note also that it is possible in principle to order genes in any manner you wish, such as ordering genes by functional category. After the identifier column, the number of subsequent columns depends on the number of measurements available for each array feature. Hence, each subsequent column represents the log ratios from a particular array. It is crucial that the ratios be log transformed (the base of the logarithm is unimportant) to ensure that positive and negative ratios are symmetric around zero. DecCor2 will work with incomplete data sets (resulting, e.g., from spots that have not passed some quality control measure), provided that empty spots in the spreadsheet are filled in using the "dummy" value of −999 (again, conveniently done with Excel). There must not be an initial header row describing the experiments.

An example file format for array data is as follows (all columns are tab-delimited):

GeneA	2.13	0.51	1.07	0.25	–2.34
GeneB	1.63	–999	5.14	–0.38	1.4
GeneC	–4.71	2.14	–999	3.23	–999

5.2.2. Gene Lists

Once an initial analysis has been performed, DecCor2 also permits the user to load a custom gene list that results in specified genes being highlighted on the gene "maps." To create a gene list, simply create a text file with the name of each unique identifier that you wish to be highlighted separated by carriage returns. Any identifiers that were not used in the original data file will be ignored, and there should be only one identifier per line.

5.3. Installing and Running DecCor2

After the program is downloaded (http://genome-www.stanford.edu/fawcett/DecCor2), clicking on the DecCor2 icon in the directory you have specified will launch the installer, which will install and optionally launch the application. Note that no other applications should be running at this time.

When DecCor2 is executed, it first prompts for the input data file name. Currently, this must be placed in the same folder as the DecCor2 executable and must be in the format specified in the previous section. As previously noted, genes in the input set can be in any desired order, but if gene names are presorted in genome order, then both the text and graphic output from the program will appear in genome order. The program will then prompt for an output file name, to be created for browsing after the user terminates the graphic output session.

DecCor2 then prompts for a "minimum vector size," corresponding to the minimum number of good data points required in a gene expression vector before DecCor2 attempts to calculate correlation coefficients for that gene. If the input set has only a few arrays, values of as small as 3 may be used, but it is preferable to use a minimum vector size of at least 10. Note that, by definition, the user must enter a number less than or equal to the number of arrays in the input data set. If the number of bad or missing data spots flagged with "–999" for a particular gene exceeds this threshold, DecCor2 concludes that insufficient data are available to generate a reliable correlation. In general, arrays representing as many different experimental conditions as possible should be included in the input set. Finally, DecCor2 will prompt the user to select a similarity metric, the options being the standard Pearson correlation coefficient or the uncentered variant presented by Eisen et al. *(7)*.

DecCor2 next parses the input file and builds a table of all the pairwise correlation coefficients using the selected similarity metric. On a slower computer, or one with less memory, this may take a significant amount of time. On an 800 MHz Pentium III with 512 megabytes of memory, it takes approx 6 min to process an input file consisting of data from 200 *E. coli* genome-size arrays (corresponding to approx 4500 rows). Note that the processing time is more sensitive to the number of genes (rows) than it is to the number of arrays (columns) in the data set, since the memory and processor time requirements grow with the square of the number of rows.

After building the correlation tables, a new display window appears showing the interactive graphic display (**Fig. 4**). This window displays a colorized map indicating the correlation coefficient between the gene name listed in the upper left-hand corner relative to all the other genes in the input data set. Presuming that the input set was in genome order, the first input gene will appear in the top left, with subsequent genes proceeding from left to right, before moving down one row at the right-hand border. Strong correlations are bright red, while strong anticorrelations are bright green. If there is no strong correlation or anticorrelation between the index gene and a map gene, the box corresponding to the map gene will have a shade approaching black. The scale bar on the left-hand side of the display indicates the correlation coefficient that corresponds to the brightest shades of red or green. This is, by default, set so that the brightest shades correspond to an absolute correlation of 1. However, this behavior can be modified using the slider on the bottom left, which effectively adjusts the contrast of the map. Adjusting the slider will adjust the numbers shown above and below the scale bar, and map genes with correlations above or below the new cutoff will be mapped to the brightest shade of red or green. Similarly, the lower right scale bar adjusts the "black point." Manipulating this slider adjusts the correlation coefficient cutoff for mapping to black those genes with minor deviations from a correlation of zero. In this way, genes with correlations or anticorrelations below those that the user considers to be significant will be mapped to black. Note that in addition to the standard red/green palette, DecCor2 supports a "heat map" style of display that may be preferred by some users; this may be accessed by checking the "brush style" box.

Rolling the mouse over the squares within the map results in a display in the top right corner indicating the name and numeric value of the correlation coefficient for the gene currently under the pointer. To switch to the correlation map for the gene under the pointer, simply left-click the mouse to redraw the map for the selected gene. Additionally, there are two other methods by which maps for other genes may be displayed. First, the scroll bar on the right-hand side of the screen can be used to select gene maps either by clicking the scroll arrows or by repositioning the scroll box. Second, the map of a gene of interest may be jumped to by typing the gene name into the search query box in the top middle of the display.

Often the user will be interested only in the behavior of a particular subset of the input genes. There are three ways that the display may be altered to accomplish this. The first method is to "filter by hits" using the filter check box. When this is checked, only those genes with the strongest gene expression profile correlations or anticorrelations will be displayed. In the map display, all other genes are displayed as gray. Genes will be shaded out if they do not deviate from the mean of the distribution of correlation coefficients by at least the amount specified by the user in the initial prompt. Note that the default distribution analyzed is the distribution of the correlations for the current gene, but, optionally, the global distribution of correlation coefficients for all genes may be used. The second method used to track the behavior of a subset of genes is to position the pointer over a gene of interest, then left double-click the mouse. This inserts a small yellow marker under the position of the gene, which allows the user to keep track of the gene as alternative maps are displayed. The user may select as many genes as desired using this method, and a second double-click on a gene results in clearing the marker for that gene. Alternatively, hitting the button labeled "clear all" simultaneously clears all such markers.

There is also provision for automatically marking a predefined set of genes, such as sets of genes that are involved in the same pathway or that have an interesting functional relation. As described under "file formats," one can create a plain-text format gene list file with each gene name of interest on a separate line. Hitting the "load" button and selecting a gene-list file automatically marks all of the specified genes. Note that genes named within the gene-list file that were not in the data input file will be ignored. Gene spellings, including capitalization, must be identical.

5.4. Concluding a DecCor2 Session

Pressing the "OK" button in the lower right-hand corner exits the graphic display portion of DecCor2 and initiates the creation of the output file. This output file will be created in the directory containing the DecCor2 executable and will have the name specified by the user during program initialization. This output file, which should be opened in a text editor such as Wordpad that supports long lines, consists of three portions:

1. A series of 200 bins containing the counts of correlation coefficients falling within each bin. Suitable for constructing a histogram of the global distribution

of correlation coefficients, this output often assumes a very normal appearance when using a large input data set.

2. A rowwise output of all the input genes, followed by a list on the same row of all the genes that had a correlation coefficient higher than the specified number of SDs above the mean of all the pairwise correlation coefficients. The cutoff correlation coefficient used for this output is the aforementioned global cutoff.

3. Another rowwise list of all genes, followed by a list on the same row of all the genes with a correlation coefficient exceeding a gene-specific cutoff. In this case, the cutoff is the mean plus the specified number of SDs calculated from the distribution of only that gene's pairwise correlation coefficients. Optionally, this list may be filtered to show only genes that are pairwise reciprocal (i.e., GeneB appears in the list for GeneA, and vice versa). Both of the gene lists generated are intended to help identify genes that are most similar in their pattern of expression to the gene listed in the first column. Note that the most similar gene is always itself, owing to the correlation coefficient of 1 for self-similarity.

References

1. Brown, P. O. and Botstein, D. (1999) Exploring the new world of the genome with DNA microarrays. *Nat. Genet.* **21(Suppl. 1),** 33–37.
2. Spellman, P. T., Sherlock, G., Zhang, M. Q., et al. (1998) Comprehensive identification of cell cycle-regulated genes of the yeast *Saccharomyces cerevisiae* by microarray hybridization. *Mol. Biol. Cell.* **9(12),** 3273–3297.
3. DeRisi, J. L., Iyer, V. R., and Brown, P. O. (1997) Exploring the metabolic and genetic control of gene expression on a genomic scale. *Science* **278(5338),** 680–686.
4. Tusher, V. G., Tibshirani, R., and Chu, G. (2001) Significance analysis of microarrays applied to the ionizing radiation response. *Proc. Natl. Acad. Sci. USA* **98(9),** 5116–5121.
5. Iyer, V. R., Horak, C. E., Scafe, C. S., Botstein, D., Snyder, M., and Brown, P. O. (2001) Genomic binding sites of the yeast cell-cycle transcription factors SBF and MBF. *Nature* **409(6819),** 533–538.
6. Lieb, J. D., Liu, X., Botstein, D., and Brown, P. O. (2001) Promoter-specific binding of Rap1 revealed by genome-wide maps of protein-DNA association. *Nat Genet.* **28(4),** 327–334.
7. Eisen, M. B., Spellman, P. T., Brown, P. O., and Botstein, D. (1998) Cluster analysis and display of genome-wide expression patterns. *Proc. Natl. Acad. Sci. USA* **95(25),** 14,863–14,868.
8. Alizadeh, A. A., Eisen, M. B., Davis, R. E., et al. (2000) Distinct types of diffuse large B-cell lymphoma identified by gene expression profiling. *Nature* **403(6769),** 503–511.
9. Pollack, J. R., Perou, C. M., Alitadeh, A. A., et al. (1999) Genome-wide analysis of DNA copy-number changes using cDNA microarrays. *Nat. Genet.* **23(1),** 41–46.
10. Hodgson, G., Hager, J. H., Volir, S., et al. (2001) Genome scanning with array CGH delineates regional alterations in mouse islet carcinomas. *Nat. Genet.* **29,** 459–464.

11. Salama, N., Guillemin, K., McDaniel, T. K., Sherlock, G., Tompkins, L., and Falkow, S. (2000) A whole-genome microarray reveals genetic diversity among *Helicobacter pylori* strains. *Proc. Natl. Acad. Sci. USA* **97(26),** 14,668–14,673.
12. Murray, A. E., Lies, D., Li, G., Nealson, K., Zhou, J., and Tiedje, J. M. (2001) DNA/DNA hybridization to microarrays reveals gene-specific differences between closely related microbial genomes. *Proc. Natl. Acad. Sci. USA* **98(17),** 9853–9858.
13. Zimmer, D. P., Soupene, E., Lee, H. L., et al., Nitrogen regulatory protein C-controlled genes of *Escherichia coli*: scavenging as a defense against nitrogen limitation. *Proc. Natl. Acad. Sci. USA* **97(26),** 14,674–14,679.

15

Microarray Databases

Storage and Retrieval of Microarray Data

Gavin Sherlock and Catherine A. Ball

1. Introduction

Microarray experiments rely on and generate *(1–3)* quantities of data that are simply too large to be stored on a researcher's desktop computer in spreadsheet format. A single microarray may have 40,000 spots. Associated with each spot may be as many as 40 different metrics, and an experimental series may consist of many dozens of such microarrays. Thus, to manage microarray data effectively, a microarray database is required; indeed, a database is not a luxury when carrying out microarray experiments—it is a necessity. In this chapter, we examine some of the properties that a microarray database should have. While this chapter focuses on the needs for tracking and storing data associated with spotted glass microarrays in particular, most of the considerations are pertinent to storage of large-scale expression data obtained on other platforms, such as Affymetrix gene chips, and nylon arrays.

It is possible to limit the information stored in a database to the information that is generated by the actual hybridization of samples to the microarray and the subsequent scanning of the microarray. However, the microarray projects usually begin much earlier than hybridizations, and the upstream data are equally important for successful analysis of microarray experiments. Thus, there are really two kinds of microarray data: those data that pertain to everything upstream of actual use of the microarray and those data from the experiment onward. The former type of data (tracking data) may be stored in a separate database, such as a laboratory information management system (LIMS), although they are needed to help one decide which experimental data are reliable, and for subsequent troubleshooting. The latter type of data, stored in the experimental database, encapsulates everything from the biological samples used and how they were treated, to the actual measurements derived from scanning the hybridized array and the normalized versions of them.

1.1. Tracking Data

1.1.1. Spotted Samples

The key piece of information that has to be recorded when using microarrays concerns the nature of the elements that are spotted on the array. Depending on exactly what is being spotted, and how it was generated, these data may be in the form of DBEST clone IDs and accession numbers, polymerase chain reaction (PCR) primer sequences, or sequences of long oligonucleotides that are spotted on the array. It has been argued that sequences should be mandatory because each clone on an array can be represented in public databases by several different IDs and accession numbers.

1.1.2. Quality Control Information from Preparation of Materials for Spotting

When either cDNA clones are spotted or specific regions of a genome are selected and amplified, they have to go through at least one round of PCR before printing. It is important that quality control information about these PCR products be tracked, because it may subsequently be useful for selection of data from a microarray database. For example, data from spots where the PCR failed, or the fragment was of an unexpected size, can be filtered out. Failed or anomalous products are printed because PCR reactions are usually done in 96-well format, so to simply spot only the products of those reactions that were deemed to have worked is impractical—instead, the content of each well in each plate is spotted. Thus, tracking the quality of those products is necessary.

1.1.3. Physical Location of Products

DNA microarrays on glass slides are usually printed by spotting DNA solutions from individual wells in the microtiter plates. Each spot on a DNA microarray uniquely corresponds to the well in a microtiter plate and not to the identity of a DNA material that has been printed. That is why it is critical to keep track of the physical location—plate coordinates—of each DNA solution. In practice, a database user will benefit from this information at least on two accounts. First, DNA replicates with the same IDs within the same or different PCR plates can be easily identified by their different locations on the printing plates. This facilitates the independent statistical analysis of hybridization data obtained on the identical DNA sequences. Second, if some nonidentical DNA sequences demonstrate very similar expression profiles, it is common practice to verify their relative location on the array. If such genes make a "streak" on the array (i.e., printed by the same print tip) located next to each other in a subarray and show identical trend in hybridization outcome ("green" or "red"), one has to rule out the possibility of cross-contamination during the print. A similar need arises when one deals with "identical expression profiles" of the genes that are situated next to each other in a print microtiter plate. Efficient recovery of the information about the print plate location of DNA solutions in question will streamline the analysis as well as help line up the control experiments.

1.2. Experimental Microarray Data

1.2.1. Biological Samples

A microarray database should allow recording of as much information as can be reliably obtained about the biological sample used in an experiment. Such information includes the organism and genotype (if known) from which the sample was derived, and the organ and/or anatomical derivation of the sample, if appropriate. In the case of human samples, a good deal of additional information could (and probably should) be added; for example, postmortem interval, cause of death, clinical diagnosis (if any) based on chart review, gender, age, medications used, toxicology screening results, and so on.

1.2.2. Protocols

During the course of a microarray experiment, many manipulations of the biological sample are likely to be carried out, and protocols for these manipulations should be faithfully recorded as part of the experiment, such that subsequent use of the data, by either the experimenters themselves or a third party, can assess exactly how and in what context the microarray data were generated. The protocols for the extraction of RNA from the biological sample, the amplification procedure, and the hybridization protocol itself should also be recorded, both for the purposes of reproducing the experiment and for troubleshooting and tracking of any protocol-dependent systematic errors that may have occurred.

1.2.3. Microarray Expression Data

Microarray expression data are the actual data generated by scanning a hybridized microarray. There are several available packages for gridding and for extracting the data from microarrays, so the database of choice should be able to support data entry from the chosen image analysis program. It is typical that the data from a microarray are not simply the ratio for each spot; there are usually multiple values per spot, such as intensity of each channel, background of each channel, and regression correlation. Many of these are useful as quality indicators of the spots. The minimal set of spot statistics that can be used for quality control usually includes the regression coefficient of the pixel-by-pixel intensities in two channels, background intensity, number of pixels at some level above background, spot size, and average and/or median intensity of the spot. However, it should be up to the individual experimenter to determine the necessary set of statistics for effective filtering without significant loss of information, which is critical in some of the downstream applications. While the normalized ratio of background-subtracted signal intensities is often all that is required to indicate the relative expression level in data analysis, one also must have an understanding of what the different spot measurements are, and how they might be used so that an informed decision can be made about which metrics to record and which to discard. Because robust analytical approaches are still in the early stages of development, it is probably wise to store as many data columns as possible, if not all of them. It is reasonable to expect that more sophisticated methods of analyzing results will be developed in the near future, so data judged unimportant today might become useful in the future. This accordingly demands greater storage space.

2. Data Retrieval

The ability to query and retrieve data effectively makes a database a powerful and often indispensible research tool. Effective retrieval, storage capacity, ease of submission, along with the flexibility and rationality of design, including the interface, are important criteria that determine the choice of a database for individual researchers.

A microarray database must allow a researcher to retrieve data about a single microarray, in a file containing all the results for the array, as well as the associated biological information. Annotation of the elements on the array is essential for interpretation of the results. Retrieval of data from multiple arrays, so that expression of genes in a group of related arrays (e.g., arrays that are part of a time series or a study of related tissue samples) can be assayed, is also a requirement of a microarray database.

2.1. Data Filtering and Extraction

A microarray database should support filtering of data during retrieval on a spot-by-spot basis, when identical sequence elements located in different wells in the print plate are treated as separate entities, as well as on a sequence ID basis, when the merged statistic (mean, median) is reported for multiple microarray entries that have the same sequence ID.

Thus, it should be possible to remove, or filter, data that are generated from spots of dubious quality, where metrics of quality may be decided by the experimenter and will depend on what data are stored within the database. Additionally, quality measurements from the manufacture of the array (whether a PCR product showed an anomaly or whether the sample might be contaminated) ought to be communicated from the LIM system (if one is used) that tracked the array synthesis. Examples of filtering metrics include flag status (indicating whether a spot was identified as being suspect, usually as the result of visual inspection of the image by the researcher, or automatic detection by image analysis software), the strength of signal to background, and the regression correlation.

For extracting information on a sequence ID basis, it should also be possible for a researcher to select data for genes that have at least a certain percentage of data present, or that vary by a certain amount in a set of arrays. Alternatively, a researcher might wish to retrieve all data for a set of genes in a group of arrays based on the identity of those genes (or at least the entities on the arrays that represent those genes), using, e.g., a list of genes of interest, or possibly selecting based on the annotation that the genes might have.

2.2. Modeling of Arrayed Samples

An easily overlooked point is how to model the entities on the arrays themselves and how they relate to the biological molecules that they represent. A simple model would be to treat the DNA sequences on the chip as if they were the genes themselves. However, this model rapidly becomes inadequate. It is obvious that this approach does not allow effective modeling of nongenic sequences (such as intergenic regions), or of large sequences that may span several genes, such as BACs. It is also ineffective at modeling gene sequences. A microarray database must model the sequences using at least two different levels of specificity: first, at the level of the physical sequence that is the entity on the microarray; and, second, at the level of the locus to which that entity maps in the genome. It is important to keep in mind when designing a microarray database that many sequences on a microarray may map to a single gene. Mapping between that sequence element on the array and a gene may evolve over time, as our knowledge of genomes and the ability to predict genes within them becomes more refined. For instance, many sequences (e.g., cDNA clones) may map to a single genetic entity, such as a Unigene cluster. The ability to retrieve data by either of these models is also useful. Retrieving clones that map to the same Unigene cluster separately is useful to be able to verify the quality of the data from the individual clones. This may allow one to determine whether a particular clone is contaminated or has been erroneously mapped to a particular Unigene cluster, if its expression pattern is different from those of other clones from the same cluster.

Alternatively, retrieval of data for multiple clones that do map to the same gene as a single, collapsed set of data is also important. Having many measurements for the same gene may subsequently affect analyses unless the many copies of the gene are downweighted in the analysis. Collapsing the data is a means to prevent this. An additional level of identification of the entities on a chip is that of instances of a piece of DNA. For example, a cDNA clone may come into the laboratory more than once, and even though each copy of the clone should ostensibly contain exactly the same sequence, it is prudent to have the ability to track these entities separately, as well as be able to collapse them to a single entity.

A more sophisticated model might map all the sequences on a microarray not just to the genes that they represent, but also to each of the exons that they may contain. Those exons could then be mapped to transcripts and the database could theoretically allow comparison of expression patterns of alternative transcripts. While this strategy is only effective with detailed knowledge of both the exons and alternative splice forms that exist, its potential value to the biologist is obvious.

2.3. Biological Annotation of Sequences

Analysis of microarray data is meaningful only in a biological context. Thus, the database holding the microarray data must also hold information that will place the data in that biological context. This can be achieved by linking the entities on the microarrays to annotation of the genes to which they map. To analyze microarray data effectively, these annotations must be accurate and up to date. The ability to achieve this is complicated by three factors. The first is that our understanding of genes and their products is changing. As more research is carried out, the information associated with many, if not all, elements on an array may be altered. The second is that the nature of annotations varies from organism to organism. For instance, the genes from one organism may be annotated by genomic coordinates and each of the three Gene Ontologies, whereas genes from another less-well-studied organism may have a simple description. Finally, for many genomes the actual mapping between a piece of DNA and the gene it is meant to represent can also change: with each release of the human Unigene clusters, several hundred clones change their cluster allegiances compared with the previous releases. Thus, a microarray database must be designed with these facts in mind and deal with them accordingly, to allow the annotation of entities on chips to be dynamic. This often means that new tables are added to the database to accommodate microarray studies on additional organisms.

2.4. Toward a Standard: MIAME

Accurate annotation of microarray experiments, to facilitate either independent verification or accurate interpretation and analysis of the generated data, is a necessity. Toward this end, members of the microarray research community are working on developing a standard, MIAME *(4)*, which defines the minimal information about microarray experiments that should be recorded. The MIAME specification, available at , is not a strict set of requirements, but really an informal set of guidelines, indicating what type of data should be recorded when microarray data are made publicly available. In the future, it is likely that journals and funding agencies will adopt such standards for data release, so when considering a database, a researcher should check whether MIAME support is either in place or intended to be developed. The MIAME specification describes experimental annotations that fall into six categories, most of which have already been discussed and are already a prerequisite of a good database:

1. *Experimental design: the set of hybridization experiments as a whole*—MIAME allows the data from a group of related microarrays to be grouped and described as a unit. Both the type of experiment can be described (e.g., a time series, a comparison of diseased tissue with healthy tissue, or comparison of a wild-type to a mutant) and the relationships of one array to another can be detailed, such as an order of the arrays within a series.
2. *Array design: each array used and each element (spot, feature) on the array*—In addition to information about the sequences spotted on the arrays, this section of MIAME records details about the platform of the microarray and production protocols. Recording this type of information has been discussed in the sections about LIM systems and modeling of sequences.
3. *Samples: samples used, extract preparation, and labeling*—The description of the samples used for hybridization should include topics such as the primary source of the biological sample; the organism from which it was derived; and the protocols used for its preparation, RNA extraction, and subsequent labeling.
4. *Hybridizations: procedures and parameters*—This section of MIAME describes the parameters and protocol of the hybridization reaction such as concentration of buffers, temperature, and length of hybridization.
5. *Measurements: images, quantitation, specifications*—The actual experimental results, progressing from raw to processed data, are covered by this section of MIAME, in the following three subsections: the original scans of the array (images); the extracted microarray data, based on image analysis; and the final data after normalization and consolidation of replicates.
6. *Normalization controls: types, values, and specifications*—There are many sources of variation in microarray experiments that affect the measured gene expression levels. Normalization is the procedure used to reduce the impact of some variations on the data generated using microarrays. To be able to compare

two or more microarray experiments, we need to remove, as best we can, the systematic variation. This section of MIAME allows recording of the scheme used for normalization, which genes may have been used for that scheme, and the algorithm that the scheme employed. In addition, the results of the normalization scheme, such as normalization factors, should be reported in this section.

3. Tools and Analysis Packages

A microarray database could simply serve as a data repository, with methods to enter and retrieve the data, as already discussed. However, it is convenient or even necessary to have various analysis and visualization tools connected directly to the database. Such tools might graph various spot parameters (e.g., channel 1 intensity vs channel 2 intensity), which may be useful for assessing overall array quality. In addition, tools for actual data analysis, such as hierarchical clustering *(5)*, self-organizing maps *(6–8)*, and principal components analysis *(9)* may come as part of a database package. Certainly, there are many commercial and free stand-alone tools for microarray analysis. If these are to be used, it is an obvious requirement that the database be able to produce data in the correct format for these tools to read.

4. Interfacing Between LIM System and Results Database

If, in addition to a microarray database, a separate LIM system is being used, then a certain amount of redundant storage between the two databases is necessary. Then the data within the LIM system that is useful for filtering microarray data can be taken advantage of. Obviously only that data required for on-the-fly filtering and analysis of the microarray results should be duplicated, since the more redundancy there is, the harder the problem becomes when data need updating. Typically, this will be the information pertaining to the plates that were used in a microarray print and the samples they contain, and then a unique identifier produced by the LIM system that tracks each instance of a piece of DNA. Interfacing between the two systems can be accomplished by production of a print file by the LIM system (sometimes referred to as a godlist), which is subsequently fed into the results database. Experiments are associated with the correct microarray print as they are entered into the database.

5. Computing Issues Associated with Microarray Databases

To use microarray data effectively, the raw data, normalized data, and all the associated spot statistics should be stored. Consequently, storing microarray data is nontrivial. While the ratio may be the salient value associated with a spot, filtering data using spot quality control metrics is an essential feature of the database. Additionally, the database should allow for an effective renormalization procedure as superior normalization techniques become available. Thus, if a typical microarray of 20,000 produces a million data points, a microarray database requires significant computing resources.

The amount of required storage is dependent on both the number of experiments that will be performed and the size of the individual arrays, and it will be greatly increased if original tiff images derived from scanning the arrays are to be archived. For example, as of August 2002, the Stanford Microarray Database (SMD) *(10)* had 28,000 experiments, whose results are recorded in approx 600,000,000 rows (averaging 21,500 rows per experiment). Physically, this takes up approx 150 gigabytes (GB), with another 80 GB needed for the indexes on this table. This storage cannot be simply accomplished using large, cheap disks, because performance would rapidly degrade as the amount of data and the number of concurrent users increase. In addition, storage of original tiff images produced by scanning a microarray requires approx 30 megabytes for each experiment (with compression). Thus, although many projects will not be generating thousands of microarrays per year, projects of all sizes must anticipate future demands and plan accordingly for the required storage.

The type of machine on which to install a microarray database is dependent on the database used, the total number of users the database will have, and how many of these users will employ the system concurrently. If the database is a multiple-user system, and many users will be retrieving and analyzing data simultaneously, then a multiprocessor machine running a Unix variant such as Solaris or Digital Unix would be strongly recommended, assuming the database management system supports it.

6. Costs Associated with a Microarray Database

Maintaining a large microarray database can be expensive. The main factors to consider in the cost are the machine(s) on which to install and house the database, the database back-end software, and the staff to maintain the database. In selecting the type of machine to house the database, the database management system and the scale of the project are the prime considerations. For the database management software itself, there are both free and commercial alternatives. The well-known commercial database management systems, such as Oracle and Sybase, will have more features and access to technical support than a free alternative such as PostgreSQL (www.postgresql.org/) and mySQL (www.mysql.com/). However, the current options for freely available database software are limited, and thus the choice of the microarray database package will probably determine what database platform is required.

7. Freely Available Databases

There are several commercial microarray databases available (*see* **ref. *13*** for details), but many of these are likely to be far beyond the budget of most laboratories. Consequently, I describe three freely available (at least to academics) databases, all of which distribute their full source code and schema, allowing customization (or improvement) by enterprising laboratories.

7.1. AMAD (Another Microarray Database)

AMAD is a flat-file database that evolved from a database that was used internally at Stanford in the Pat Brown and David Botstein laboratories and was written by Mike Eisen, Paul Spellman, Joe DeRisi, and Max Diehn. AMAD is available from http://microarrays.org/software.html.

7.1.1. Requirements and Installation

AMAD requires a computer that can host a Web server and run Perl, such as a PC running Linux. Installation is fairly simple, and there is a detailed installation document explaining how to do it.

7.1.2. Features

AMAD allows data uploading and subsequent viewing and retrieval of that data. Experimental data can be viewed on an experiment-by-experiment basis, or data can be retrieved for multiple experiments simultaneously, into a single file, for use in subsequent analyses (output files are compatible with Mike Eisen's Cluster software). Filtering of the data by any column of data that is present in the uploaded files may be done when retrieving data from AMAD, and the user can add additional arbitrary columns of data when defining the contents of a microarray print, and AMAD will transfer that data to the experimental data files. AMAD supports loading of data from both Scanalyze (http://rana.lbl.gov/EisenSoftware.htm) and GenePix (www.axon.com/GN_GenePixSoftware.html), two commonly used software packages for analyzing scanned microarray images.

7.1.3. Advantages

AMAD is a simple and effective database for a laboratory with a modest budget and with a few dozen to a few hundred arrays' worth of data. It is easy to install and has an uncluttered, intuitive interface.

7.1.4. Disadvantages

AMAD is not a relational database and thus does not have many of the features that a relational database provides. Consequently, there is a large amount of data redundancy (e.g., every experiment coming from a given print reproduces all of the same print information), which means that when annotation or tracking information needs updating, many files may require modification. The limitations of the flat-file system make this package impractical for any group of investigators who expect to carry out a substantial number of microarray experiments. In addition, there is no simple way to update an experiment if an error in the print list is made, other than reentering the experiment with a corrected print list. AMAD is no longer under active development and, consequently, will not be MIAME compliant—a replacement, NOMAD, which uses the free relational database mySQL, is under development.

7.2. Stanford Microarray Database

The SMD uses Oracle as its database management system and is thus a relational database. Its full source code and schema are available from http://smd.stanford.edu.

7.2.1. Requirements and Installation

The Stanford installation of SMD is on a Sun E4500 server running Solaris 2.8 with four processors and 8 GB of random access memory. While there is no specific reason to use a Solaris system, Oracle updates and bug fixes often appear first on this platform, making it desirable. SMD installation requires Oracle Enterprise Edition server software, a Web server, Perl, and several Perl modules. This is currently not simple, but an installer script distributed with the software does take care of many of the details of getting it running. Additional details, such as setting up the Oracle instance of the database and creating all the tables and the relationships among them, do require a trained database administrator, although all the SQL scripts required to do this are distributed with the SMD package.

7.2.2. Features

SMD allows entry of either GenePix or Scanalyze data (in either single or batch mode) and subsequent retrieval of data from one or many experiments, with complex filtering, in much the same way as AMAD does. In addition, SMD has many tools built on top of the database that may be used for assessing array quality and experimental reproducibility, visualizing a representation of the original microarray, and performing downstream analyses. The latter are currently limited to hierarchical clustering and self-organizing maps, but future plans are to encompass *K*-means clustering *(11)*, singular value decomposition *(9)*, and imputation of missing data *(12)*.

7.2.3. Advantages

SMD is a scaleable solution for storing microarray data—the Stanford installation currently has >30,000 experiments, constituting data from approx 600,000,000 spots—while a flexible security model allows fine-grained access control to both data and tools. Several tools are available with the database, and software for viewing proxy images of the microarray scans to visually evaluate the quality of the data are also available. Furthermore, SMD dynamically updates annotation of the human, mouse, and yeast genes that are represented on the microarrays and is under active development, so that new features and improved schema and software will become available in the future.

7.2.4. Disadvantages

SMD requires expensive hardware and software, as well as trained staff (at least a database administrator and a programmer/curator) to keep it running, and it is not simple to install or maintain. Currently, SMD does not store enough experimental or sample information to comply with the MIAME specifications, but this problem should be corrected by early 2003. SMD was designed to store only two-channel microarray data and thus will not store Affymetrix data or nylon membrane data.

7.3. GeneX

GeneX (http://genex.ncgr.org) is a freely available database (released under the GNU lesser public license to academics) that uses the free database system PostgreSQL to store the data.

7.3.1. Requirements and Installation

GeneX can be set up on a Linux machine running an Apache Web server with Perl installed. An installation script is used to configure the components of the system.

7.3.2. Features

A client-side curation tool, written in Java, is provided for formatting data sets for secure upload into the database, and then simple html interfaces for retrieving that data either across data sets or for a single experiment can be used. In addition, the software that ships with GeneX provides an application programming interface that allows experienced programmers to make extensions to the interface. There are a number of analytical routines that can be run against the retrieved data, such as clustering, multidimensional scaling, cluster validation, and principal component analysis, that are provided as separately available add-ons for the database.

7.3.3. Advantages

Like AMAD, GeneX is fairly simple to install and does not require expensive hardware or software to install or maintain. It has a well-considered data model and is also able to store data from different array platforms, such as two-color microarray data and single-channel Affymetrix data. It has a number of available tools with which to analyze data. GeneX is under active development so that one can expect improvements to the software, data modeling, and analysis tools, and it has already had several small bug fix updates since its initial release early in 2001.

7.3.4. Disadvantages

GeneX has not yet been demonstrated to scale to hold data for many thousands of experiments, although that, of course, does not mean that it will not do so. GeneX does not allow viewing of proxy images for characterization of array quality.

8. Conclusion

A microarray database is a necessity for microarray research, but setting up and maintaining such a database is no simple proposition, potentially requiring significant computing resources and trained personnel. A number of free databases are available for local installation, but none of these alternatives is mature enough to offer simple installation and meet all users' needs. Of course, there are many available commercial databases (*see* **ref. *13***), but the costs of initial purchase (and often subsequent annual licensing fees) are prohibitive to all but the largest of laboratories. Before embarking on a microarray project, a researcher should carefully consider his or her needs regarding a database—indeed, when applying for funding this should be an important consideration. Without an adequate database, microarray projects are surely doomed to failure, and database needs should be planned just as meticulously as the experimental results that the database will eventually house.

References

1. Lockhart, D. J., Dong, H., Byrne, M. C., Follettie, M. T., Gallo, M. V., Chee, M. S., Mittmann, M., Wang, C., Kobayashi, M., Horton, H., and Brown, E. L. (1996) Expression monitoring by hybridization to high-density oligonucleotide arrays. *Nat. Biotechnol.* **14(13),** 1675–1680.
2. DeRisi, J. L., Iyer, V. R., and Brown, P. O. (1997) Exploring the metabolic and genetic control of gene expression on a genomic scale. *Science* **278(5338),** 680–686.
3. Schena, M., Shalon, D., Heller, R., Chai, A., Brown, P. O., and Davis, R. W. (1996) Parallel human genome analysis: microarray-based expression monitoring of 1000 genes. *Proc. Natl. Acad. Sci. USA* **93(20),** 10,614–10,619.
4. Brazma, A., Hingcamp, P., Quackenbush, J., et al. (2001) Minimum information about a microarray experiment (MIAME)—toward standards for microarray data. *Nat. Genet.* **29(4),** 365–371.

5. Eisen, M. B., Spellman, P. T., Brown, P. O., and Botstein, D. (1998) Cluster analysis and display of genome-wide expression patterns. *Proc. Natl. Acad. Sci. USA* **95(25)**, 14,863–14,868.
6. Kohonen, T. (1995) *Self Organizing Maps*, Springer, Berlin.
7. Tamayo, P., Slonim, D., Mesirov, J., Zhu, Q., Kitareewan, S., Dmitrovsky, E., Lander, E. S., and Golub, T. R. (1999) Interpreting patterns of gene expression with self-organizing maps: methods and application to hematopoietic differentiation. *Proc. Natl. Acad. Sci. USA* **96(6)**, 2907–2912.
8. Toronen, P., Kolehmainen, M., Wong, G., and Castren, E. (1999) Analysis of gene expression data using self-organizing maps. *FEBS Lett.* **451(2)**, 142–146.
9. Alter, O., Brown, P. O., and Botstein, D. (2000) Singular value decomposition for genome-wide expression data processing and modeling. *Proc. Natl. Acad. Sci. USA* **97(18)**, 10,101–10,106.
10. Sherlock, G., Hernandez-Boussard, T., Kasarskis, A., et al. (2001) The Stanford microarray database. *Nucleic Acids Res.* **29(1)**, 152–155.
11. Everitt, B. (1974) *Cluster Analysis*, Heinemann, London.
12. Troyanskaya, O., Cantor, M., Sherlock, G., Brown, P., Hastie, T., Tibshirani, R., Botstein, D., and Altman, R. B. (2001) Missing value estimation methods for DNA microarrays. *Bioinformatics* **17(6)**, 520–525.
13. Gardiner-Garden, M. and Littlejohn, T. G. (2001) A comparison of microarray databases. *Brief Bioinform.* **2(2)**, 143–158.

Index

A

Alternating design, 143, 144
AMAD database,
  advantages, 244
  disadvantages, 244, 245
  features, 244
  requirements and installation, 244
Arraying, *see* Fabrication, cDNA microarrays
Atlas Plastic Film,
  advantages, 31, 32
  binding capacity and efficiency, 33
  calibration, 35
  hybridization,
    incubation conditions, 48–50, 52
    reagents, 39
    time and efficiency, 33
  normalization of signals, 51, 52
  printing,
    arrayer, 40
    crosslinking, 40, 41
    mix preparation, 40
    overview, 39, 40
    protective sheet removal, 40, 51
    reagents, 35–37
  probe synthesis,
    cDNA probe synthesis, 47, 48, 52
    poly(A) RNA enrichment, 46, 47, 52
    streptavidin magnetic bead preparation, 46
  resolution, 50, 51, 53
  RNA isolation,
    Atlas Pure Total RNA Labeling System, 37, 38
    cultured cell RNA isolation, 42–44, 52
    DNA contamination testing with polymerase chain reaction, 45, 46
    DNase treatment, 44, 45
    gel electrophoresis, 45
    reagents, 37–39, 41, 43
    spectrophotometric quantification, 45
    tissue RNA isolation, 41, 42, 52
    yields, 44
  sensitivity, 33, 35, 50, 51, 53
  stripping, 39, 50, 52

B

Background,
  array quality, 123
  estimation approaches, 115, 116
  nylon membrane levels, 25, 26
  subtraction, 116
Box plot, graphic presentation of slide data, 118, 119
B statistic, gene ranking, 127, 128, 156

C

Caryoscope, microarray data visualization, 225

Chromatin immunoprecipitation, protein–DNA interaction mapping,
  chromatin immunoprecipitation, 103, 108
  crosslinking,
    chromatin complex, 101, 102, 108
    reversal reaction, 103
  DNA microarray analysis,
    data analysis, 107–109
    DNA purification, 103, 104, 108
    fluorescent labeling, 105
    polymerase chain reaction, 104, 105
    probe purification and hybridization, 105–107
  extract preparation, 102, 108
  materials, 100, 101, 108
  principles of mapping with DNA microarrays, 99, 100
  yeast culture, 101
Class discovery, definition, 131
Class prediction, approaches, 131, 132
Clustergram, data presentation, 222
Clustering,
  algorithms,
    feature-based algorithms, 186, 187
    globular clustering algorithms, 187, 188
    hierarchical clustering algorithms, 163, 164, 184–186
    partitional algorithms, 164, 184, 185
    selection considerations, 165
    similarity-based algorithms, 186
    similarity versus dissimilarity matrix, 162
    transitive clustering algorithms, 187, 188
  applications, 183, 184
  associating clusters of two partitions, 168–170
  class prediction, 131, 132
  CLUTO software,
    algorithm overview, 203–205, 207, 208
    availability, 203
    cluster analysis features, 208, 211
    clustering criterion function, 203, 204
    graph-partitioning-based clustering, 204, 205
    parameters for clustering algorithms, 206, 207
    tree-building for large data sets, 208
    usage, 205, 206
    visualization of clusters, 211, 213
  coregulated genes, 199
  group average approach, 185
  inducer grouping, 199
  K-means approach, 202, 203
  meaningful groups, 183
  principal components analysis, stability-based validation of cluster structure,
    advantages, 179, 180
    centering, 161, 162
    notation, 161, 162
    overview, 160
    principal components definition, 161
    standardization, 162
    total variance selection limitations, 178

yeast gene expression data example,
    centered variable clustering, 170–172, 174, 175
    functional classes of genes, 170
    Pearson correlation similarity measure, 178
    stability-based choice of principal components numbers, 175, 176
    standardized variable clustering, 176–178
principles, 159
quality assessment of clusters, 196–199
self-organizing maps, 202, 203
similarity measurements between objects,
    multidimensional objects,
        categorical attributes, 192, 193
        continuous attributes, 190–192
        overview, 189
    sequences,
        global sequence alignment, 193, 194
        heuristic approaches, 194
        local sequence alignment, 194
        row similarity scores, 194, 195
        types of objects, 189
    structures,
        proteins, 195, 196
        root mean square deviations, 195
        superimposition, 195, 196
        types of objects, 189
similarity measures in microarrays, 201, 202
stability, 165–168

subspace, 187
tissue type, gene expression grouping, 199
CLUTO, *see* Clustering
Copy number, *see* Gene copy number, cDNA microarray analysis

D

Databases, microarrays,
    AMAD database,
        advantages, 244
        disadvantages, 244, 245
        features, 244
        requirements and installation, 244
    analysis package interface, 242
    commercial databases, 243, 244, 247
    costs, 243
    data,
        complexity, 235
        types, 235, 236
    Escherichia coli, *see Escherichia coli* DNA microarrays
    experimental data,
        biological samples, 237
        expression data, 237, 238
        protocols, 237
    GeneX database,
        advantages, 246, 249
        disadvantages, 247
        features, 246
        requirements and installation, 246
    hardware requirements, 243
    laboratory information management system interface, 242
    retrieval of data,
        data filtering and extraction, 238, 239

MIAME specifications, vii, 241, 242
modeling of biological samples, 239, 240
overview, 238
sequence annotation, 240
Stanford Microarray Database,
  advantages, 245, 246
  disadvantages, 246
  features, 245
  requirements and installation, 245
storage requirements, 243
tracking data,
  physical location of products, 236, 237
  quality control information, 236
  spotted samples, 236
DecCor2
  data formats,
    array data, 229, 230
    gene lists, 230
  hardware requirements, 229
  installation and execution, 230–232
  outlier handling, 228
  output file, 232, 233
  overview, 225, 227
Double loop design, 143, 144
Double reference design, 143, 144
Double-strand DNA microarrays, *see Escherichia coli* DNA microarrays
Dye bias, experimental layout, 144, 145

E

Efficiency, experimental layout, 141–144
Escherichia coli DNA microarrays,
  applications, 61
  array preparation, 71
  cDNA synthesis and fluorescent labeling,
    direct labeling,
      annealing, 69, 76
      cleanup, 69, 76
      materials, 63
      reverse transcriptase reaction, 69, 76
    indirect labeling,
      annealing, 67
      cleanup, 67, 68, 76
      coupling reaction, 68
      materials, 63
      reverse transcriptase reaction, 67, 76
      RNA hydrolysis, 67
      unincorporated dye removal, 68, 69
  formaldehyde gel denaturation, 62
  hybridization, 63, 64, 71
  Internet resources,
    Colibri, 74
    EcoCyc, 75
    EcoGene, 74
    E. coli Entry Point, 75
    GenProtEC, 75
    NCBI, 74
    RegulonDB, 75
  normalization, 72, 77
  quality control, 72, 73
  RNA isolation,
    DNase treatment, 62, 66
    ethanol precipitation, 66
    materials, 61, 62
    mRNA,

Index 253

cell culture and harvesting, 65, 76
cell lysis, 65, 76
enrichment, 66, 76
materials, 62
phenol/chloroform extractions, 66
quality assessment, 67
RNase avoidance, 76
total RNA,
  cell culture and harvesting, 64, 76
  cell lysis, 64
  chloroform extraction, 65
  ethanol precipitation, 65
  hot phenol extraction, 64, 65
  materials, 61, 62
sample preparation and application, 70, 71
washing, 71, 72, 76
Experimental design, microarray analysis,
  comparison of similarly-designed experiments, 147
  layout,
    dye bias, 144, 145
    efficiency, 141–144
    extendibility, 146
    overview, 141
    robustness, 145
    simplicity, 145, 146
    useful subdesigns, 146
  pooling, 139, 140
  randomization, 146
  replication, 138, 139, 146
  statistical analysis issues, 112–114

variables, 137
Extendibility, experimental layout, 146

F

Fabrication, cDNA microarrays,
  glass slides, *see* Glass slides, cDNA microarrays
  nylon arrays, *see* Nylon cDNA expression arrays
  plastic arrays, *see* Atlas Plastic Film
  spots,
    size, 1
    spacing, 1
False discovery rate (FDR),
  estimation, *see* Significant analysis of microarrays
  gene ranking, 150–152
  minimum positive value, 155
Familywise error rate (FWER), gene ranking, 150, 151, 156
FDR, *see* False discovery rate
Fluorescent probes,
  cDNA,
    dye coupling, 58
    purification, 57, 58
    RNA removal, 57
    synthesis, 57
  chromatin-immunoprecipitated DNA labeling, 105
  Escherichia coli, see Escherichia coli DNA microarrays
  graphic presentation of slide data, 117–120
  hybridization, 58, 59
  materials for preparation, 56
  principles of preparation, 55, 56

purification, 58
RNA,
  isolation for cDNA synthesis, 57
  quantity requirements, 55, 56
  washing, 58, 59
FWER, *see* Familywise error rate

G

GeneChip, principles, 200
Gene copy number, cDNA microarray analysis,
  advantages, 91
  applications, 89, 91
  data reduction and analysis, 93, 95
  genomic DNA,
    hybridization, 91, 92, 94, 95
    isolation, 93, 95
    labeling, 91, 93–95
  materials, 91–93
  microarray preparation, 93
  principles of parallel analysis, 89, 90
  RNA labeling and hybridization, 93
  washing, 94, 96
GeneX database,
  advantages, 246, 249
  disadvantages, 247
  features, 246
  requirements and installation, 246
Glass slides, cDNA microarrays,
  advantages, 1
  clone selection and preparation, 2, 3
  limitations, 31
  materials, 2
  plasmid purification, 3
  poly-L-lysine coating, 4, 5
  polymerase chain reaction, 3, 4
  postprocessing, 6
  principles, 199, 200
  robotic spotting, 5, 6

H

Hierarchical clustering, algorithms, 163, 164, 184–186

K

K-means approach, clustering, 202, 203

L

Laboratory information management system (LIMS), database interface, 242
LIMS, *see* Laboratory information management system

M

M, mean values for gene ranking, 126, 127, 129
MIAME, microarray data specifications, vii, 241, 242
Microarray sample pool (MSP), titration series for normalization, 122
MSP, *see* Microarray sample pool

N

Normalization,
  Atlas Plastic Film signals, 51, 52
  Escherichia coli DNA microarrays, 72, 77
  global constant, 120, 121
  importance, 120
  intra- versus interarray, 120
  invariant gene utilization, 122, 123
  microarray sample pool titration series, 122
  nylon cDNA expression arrays, 26, 27

principles, 201
print-tip normalization, 121, 122
Nylon cDNA expression arrays,
  advantages, 9
  background level, 25, 26
  cDNA,
    polymerase chain reaction, 14, 15
    purification, 14, 15
    standardization, 15
  exposure time, 26
  hybridization, 13, 24, 27
  limitations, 31
  normalization of signals, 26, 27
  printing of membranes,
    arrayer, 16
    controls, 15, 16
    crosslinking, 16
    equipment, 11, 12
    mix preparation, 16
    reagents, 11
  probe synthesis,
    cDNA probe synthesis, 22, 23, 27
    poly(A) RNA enrichment, 22
    streptavidin magnetic bead
      preparation, 21, 22
  RNA isolation,
    Atlas Pure Total RNA
      Labeling System, 12
    cultured cell RNA isolation,
      18–20, 27
    DNA contamination testing
      with polymerase chain
      reaction, 21
    DNase treatment, 20, 21
    gel electrophoresis, 21, 27
    reagents, 12, 13
    spectrophotometric
      quantification, 21
  tissue RNA isolation, 16–18, 27
  sensitivity, 9–11, 25, 26
  stripping, 13, 25, 27, 28
  washing, 13, 24, 25

P

PCA, *see* Principal components
  analysis
PCR, *see* Polymerase chain reaction
Photomultiplier tube (PMT),
  normalization, 121
  voltage and sensitivity, 114
Pixel,
  intensity scale, 123
  threshold, 114, 115
Plasmid, purification for glass slide
  cDNA microarrays, 3
Plastic arrays, *see* Atlas Plastic Film
PMT, *see* Photomultiplier tube
Polymerase chain reaction (PCR),
  cDNA for nylon cDNA
    expression arrays, 14, 15
  DNA contamination testing of
    RNA, 21, 45, 46
  glass slide cDNA microarray
    preparation, 3, 4
Polysomal RNA microarray
  analysis,
  applications, 79, 80
  cell lysis,
    mammalian cells, 83, 86
    yeast, 82, 83, 86
  materials, 81, 82, 86
  polysomal profile, 79
  RNA labeling and hybridization, 85
  sucrose gradient,
    fraction collection, 83, 84, 86
    preparation, 82, 86

RNA extraction, 84, 85
Pooling, RNA from many individuals, 139, 140
Principal components analysis (PCA), stability-based validation of cluster structure,
  advantages, 179, 180
  centering, 161, 162
  notation, 161, 162
  overview, 160
  principal components definition, 161
  standardization, 162
  total variance selection limitations, 178
  yeast gene expression data example,
    centered variable clustering, 170–172, 174, 175
    functional classes of genes, 170
    Pearson correlation similarity measure, 178
    stability-based choice of principal components numbers, 175, 176
    standardized variable clustering, 176–178
Promoter, microarray data visualization, 223
Protein–DNA interaction mapping, *see* Chromatin immunoprecipitation, protein–DNA interaction mapping
p value, gene ranking, 130, 155, 156

Q

q value, gene ranking, 155

R

Randomization, experimental design, 146

Reference design, 142, 143
Replication,
  experimental design, 138, 139, 146
  subsampling, 138
  types, 138
  variability in microarray experiments, 139
RNA,
  isolation,
    Atlas Pure Total RNA Labeling System, 12, 37, 38
    cultured cell RNA isolation, 18–20, 27, 42–44, 52
    DNA contamination testing with polymerase chain reaction, 21, 45, 46
    DNase treatment, 20, 21, 44, 45
    Escherichia coli, see Escherichia coli DNA microarrays
    fluorescent probe preparation, 57
    gel electrophoresis, 21, 27, 45
    polysomal RNA, *see* Polysomal RNA microarray analysis
    reagents, 12, 13, 37–39, 41, 43
    spectrophotometric quantification, 21, 45
    tissue RNA isolation, 16–18, 27, 41, 42, 52
    yields, 44
  probe synthesis,
    cDNA probe synthesis, 22, 23, 27, 47, 48, 52
    Escherichia coli, see Escherichia coli DNA microarrays

Index

poly(A) RNA enrichment, 22, 46, 47, 52
streptavidin magnetic bead preparation, 21, 22, 46
Robustness, experimental layout, 145

S

SAM, *see* Significant analysis of microarrays
Scatter plot, graphic presentation of slide data, 117, 118
Segmentation, techniques, 115
Self-organizing map (SOM), clustering analysis, 202, 203
Signal-to-noise ratio (SNR), spot quality assessment, 124–126
Significant analysis of microarrays (SAM),
 false discovery rate estimation, 153–156
 software availability, 156
Simplicity, experimental layout, 145, 146
Single loop design, 142, 143
SMD, *see* Stanford Microarray Database
SNR, *see* Signal-to-noise ratio
SOM, *see* Self-organizing map
Spatial plot, graphic presentation of slide data, 118–120
Spot quality, assessment, 124–126
Spotting, *see* Fabrication, cDNA microarrays
Stanford Microarray Database (SMD),
 advantages, 245, 246
 disadvantages, 246

features, 245
requirements and installation, 245
Statistical analysis, cDNA microarrays,
 array quality, 123, 124
 Bayesian method for gene ranking, 127, 128, 156
 challenges, 149, 150
 class prediction and discovery, 131, 132
 clustering, *see* Clustering
 differentially expressed gene selection,
  ranking genes, 126–129
  significance assignment, 129–131
 experimental design issues, 112–114
 false discovery rate, 150–152
 false positives, 149
 familywise error rate, 150, 151, 156
 graphic presentation, 117–120
 image analysis issues, 114–117
 multiple hypothesis testing, 149, 150
 normalization, 120–123
 overview, 111, 112, 133
 pooled standard deviation, 150
 power, 149
 p value, 130, 155, 156
 q value, 155
 rank-sum statistic, 150
 significant analysis of microarrays, 153–156
 software, 133
 spot quality, 124–126
 t statistic, gene ranking, 126–129, 150, 152
Sucrose gradient, polysomal RNA microarray analysis,
 fraction collection, 83, 84, 86
 preparation, 82, 86

RNA extraction, 84, 85
Symmetric reference design, 143, 144

T

TreeView, microarray data visualization, 220, 222, 223
t statistic, gene ranking, 126–129, 150, 152

V

Visualization, microarray data,
  box plot, 118, 119
  Caryoscope, 225
  challenges, 219
  CLUTO software visualization of clusters, 211, 213
  DecCor2
    data formats,
      array data, 229, 230
      gene lists, 230
    hardware requirements, 229
    installation and execution, 230–232
    outlier handling, 228
    output file, 232, 233
    overview, 225, 227
  expression ratio representation, 220
  Promoter, 223
  scatter plot, 117, 118
  spatial plot, 118–120
  TreeView, 220, 222, 223